SpringerBriefs in Economics

More information about this series at http://www.springer.com/series/8876

Stefan Mann

Socioeconomics of Agriculture

Stefan Mann
Socioeconomics
Agroscope
Ettenhausen
Switzerland

ISSN 2191-5504   ISSN 2191-5512   (electronic)
SpringerBriefs in Economics
ISBN 978-3-319-74140-6   ISBN 978-3-319-74141-3   (eBook)
https://doi.org/10.1007/978-3-319-74141-3

Library of Congress Control Number: 2017963857

Printed on acid-free paper

This Springer imprint is published by the registered company Springer International Publishing AG part of Springer Nature
The registered company address is: Gewerbestrasse 11, 6330 Cham, Switzerland

Contents

Chapter 1
What Is This Book Good for?

Socioeconomics may be under greater pressure to define itself than the "classical" sciences. The latter have largely been defined by the conceptualization of real-world phenomena: biology, for example, arose because of the willingness to better understand the phenomenon of life in its different specificities. Economics arose to study and resolve the issue of scarcity. And sociology was created to analyze the dynamics of societal groups.

But socioeconomics? Its name pays tribute to the existence of sociology and economics, so one might either suspect that socioeconomics is merely is a combination of both sciences (S + E = SE), or that there is an intersection of sociology and economics which is best termed as socioeconomics (S ∩ E = SE). Most past attempts to define socioeconomics as a science in its own right may have been motivated to counter such a simplistic understanding of socioeconomics.

In this chapter, we review past attempts to define socioeconomics before the approach is chosen that we applied in this book.

1.1 Approaches Towards Socioeconomics

There is a strong lingual proximity between socioeconomics and social economics. Social economics, however, is a discipline with a considerable tradition. Since its early beginnings (Ward 1893), proponents of social economics argue that economics is not just the replication of a biological system in society, but that moral and social considerations always have to be considered. Social economic was called everything that attempted to improve the living standard of the working class (Say 1891; Gide 1905). This has hardly changed to date. The Association of Social Economics claims on its website that "social economics is the study of the ethical and social causes and consequences of economic behavior, institutions, organizations, theory, and policy". It probably would not be necessary to mention that this association had been founded in 1941 as the "Catholic Economic Association" to convince readers

© The Author(s) 2018
S. Mann, *Socioeconomics of Agriculture*, SpringerBriefs in Economics,
https://doi.org/10.1007/978-3-319-74141-3_1

that social economics follows a rather normative rationale. It goes beyond the utilitarian fundamentals of economic science and emphasizes moral values in societal and business decision-making.

Whatever can be said about socioeconomics, it is clear that it takes a much more descriptive approach than social economics, even though it is also considered as alternative to mainstream economics. Socioeconomics also is a lot younger than social economics. Since the 1970s, the word was occasionally used in the non-economic literature to describe developments touching both sociological and economic aspects. However, the leading communitarian scholar Etzioni (1985, 1986) was the first prominent voice to suggest that "a new discipline would be developing which would merge economics and other social sciences" (Etzioni 1986: 13). He suggested four fields in which such an approach would generate added value:

> (1) opening up the preferences; (2) modifying the assumption of rationality (again!) (3) the societal nestling of the market (a matter of institutions and political power); and (4) increasing the empirical, inductive elements of the study of economic behavior. (Etzioni 1986: 13)

The impulse which Etzioni gave was strong enough that the Society for the Advancement of Socio-Economics (SASE) was founded in 1988 and that the Journal of behavioral economics, in 1990, was re-named into the Journal of Socio-Economics (23 years later, though, it was then re-renamed into the Journal of behavioral and experimental economics).

Since then, both the journals and the conference have flourished and attracted increasing number of submitters. Particular empiric contributions are apparently attractive to make, ranging from "Are High-Performance HR Practices Good for Employee Well-being" to "How Race and Human Resources Influence Consumer Expectations and Attitudes". Theoretic contributions about the nature of socioeconomics are somewhat rarer. However, a few of the brave colleagues having made attempts to proceed on the theoretical side of the socioeconomic agenda are certainly worth mentioning:

– Abell (2003) distinguishes two dimensions relevant for sociology, economics and (in his argument) socioeconomics. One are interactions on the micro level versus social (macro) conditions, the other dimension is whether to focus on the interactions and conditions themselves, or to look at the individual actions or the social outcome. Abler wants to "unite" sociologists and economists by suggestion to focus on the causalities of individual interactions and social conditions and to analyze how both impact on individual actions. He emphasizes the different starting points that economics and sociology have regarding models of the individual. In the economic view, "individuals are conceived as taking choices" (Abell 2003: 8), so they are considered as subjects, whereas for many sociologists "individuals are [...] deriving their actions (or decision to act) from those with whom they interact" (Abell 2003: 9), so they are more considered as objects.
– Karl H. Müller (Hollingsworth and Müller 2008; Müller 2015, 2016) has a historic approach, summarizing the classic, reductionist way of research as in classic physics as "science 1" and the modern, complex approach as found in life sciences

as "science 2". He sees the challenge for socioeconomists to enter the "science 2" mode by occupying themselves with complex networks, evolutionary theory and nanofoundations as a significant degree below the micro-level and by deliberately entering more meta studies to uncover the limitations of researchers themselves. Much more than anybody else, Müller tries to advance rather than define socioeconomics, i.e. his priority lies in the progress of the socioeconomic discourse rather than in the establishment of a clearly defined socioeconomics as an additional social science.

– In a critique to the last approach, Boyer (2008: 744) suggests that "socio-economics is about the investigation of the origin, transformation and impact of governance structures in modern societies". Socioeconomics should focus on the institutional arrangements people have given themselves to organize social, economic and political relations. Such analyses could result in formalizations of the models identified. It could also result in the identification of viable institutional settings on the macro level. His vision of socioeconomics also includes the added value from comparative historical institutional analysis.

This book certainly owes most of its epistemologic foundations to Boyer's ideas. How the latter might be translated into practice and will be translated in this book, is to be outlined in the remainder of this chapter.

1.2 The Interaction Approach

Most institutional economists subscribe to the notion that three different modes of interaction should be distinguished: Since Adam Smith's "The Wealth of Nations" had established the science of economics in 1776, the discipline's focus has increasingly been put on trade and on markets. The economic historian Mikl-Horke (2015) nicely illustrates this reduction of economic science, particularly in the decades around 1900. While early economists would integrate social, political and ethic aspects in their "economic" thinking, economics was increasingly reduced in the late 19th century to restrict itself to the exchange of goods, leading to a (mostly implicit) world-view in which everything was subject to trade. It then was Coase (1937) who related the discourse on interaction to the existence of companies and their hierarchies. He posed the question why such (sometimes huge) hierarchies would be created if everything would be most efficiently traded on markets, thus demonstrating some institutional diversity. Table 1.1 demonstrates the size of the hierarchies by comparing the turnover of the largest private companies to the GDP of countries.

Coase remarked that defining prices would cause transaction costs, as do hierarchies in their daily routine. Coase suggested that there would be a social optimum somewhere in between a centrally planned world hierarchy and atomized trading partners. While enterprises, in classical economics, largely had been defined as a production function, they now became a structural element in a society's organization.

Table 1.1 Turnover of private companies and GDP of comparable countries (2016)

Private company	Turnover (billion US-$)	Country	GDP (billion US-$)
Wal-Mart	482	Belgium	470
Royal Dutch Shell	272	Pakistan	271
Volkswagen	237	Finland	239

Institutional economists have devoted a lot of energy towards the question of why persons would choose either hierarchies or markets as coordinating mechanism of their interactions. Three important factors—asset specificity, frequency and uncertainties—have been identified in the wake of this research which had a strong predictive power on the level of transaction costs in this context and therefore would also influence organizational choices:

- If the specificity of assets is high, this creates a strong mutual dependency which suggests a long-term hierarchical organization rather than short-term trade on markets. Courses in milking technology will rather be organized in an intra-firm setting than courses in presentations.
- The more often a transaction takes place, the more transaction costs can be distributed. Frequent transactions may rather be organized within one unit and in a hierarchical framework, so that repeated negotiations may be avoided.
- Uncertainty increases transaction costs, so that long-term hierarchies may be chosen to avoid such uncertainties.

The advantages of hierarchies are that human resources can rather flexibly be reallocated, mental work can be distributed and shared more flexibly than in a pure market setting and a well-established communication system is also more easily established within a hierarchy.

More than half a century later than Ronald Coase, Ostrom (1990) introduced coordination as another important mode of interaction, showing that often people chose to jointly work for a common goal by neither trading nor submitting to each other. Everyday examples for cooperation are traffic, sports or associations.

If we think about the most beautiful incident of our life, it is likely that we find it in the realm of cooperation, be it a romantic encounter, an occasion with our friends or joint work. While economists love to talk about utility maximization, they tend to neglect the fact that it is unlikely that individual utility maximizations, added up, will lead to a societal optimum by itself. It is true that, as economists often emphasize, cooperation suffers from possible opportunistic behavior like the free-rider phenomenon. It is often possible to exploit a team to which I become a member by contributing less than benefitting, sometimes to the point of the collapse of the system. The spirit that joint and mutual actions produce, however, is nevertheless driving many people to, again and again, enter attempts of successful cooperation Fig. 1.1.

Having mentioned cooperation and hierarchies as important occasions of interaction, they have never made it into the core domain of economic analysis. It is no

Fig. 1.1 Three modes of interaction

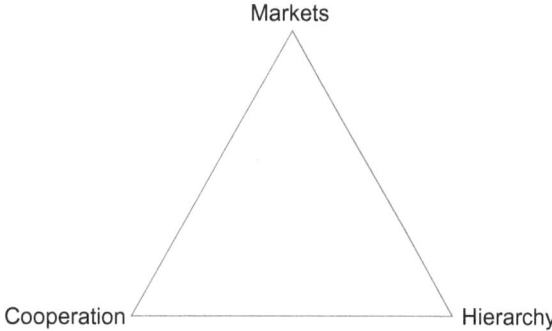

coincidence that Ronald Coase has always been a fierce critic of mainstream economics (Coase 1946, 1988; Schwab 1989), and that Eleanor Ostrom was not even an economist by education (but was rather a political scientist). Neither is it coincidence that the concepts of hierarchy and cooperation can rarely be found in the mainstream textbooks and concepts of economists. The assumptions of markets are an ideal playing field for the reductionist and quantitative way of working that most economists feel comfortable with.

This is not to say that cooperation and hierarchy would not be apt for scientific analysis. A great part of political science, for example, deals explicitly with the mechanics of hierarchy (Lieshout 1995; Wight 2002), both its creation (Pabst 2012) and its impact (Ikeda and Richey 2005). The same can be said about management science (Magee and Galinsky 2008; Friesl et al. 2011), albeit for private rather than for public environments. Grey (2004) explores the relations between both realms of research and their strong link via the analysis of hierarchies.

Anthropology is the science most concerned with cooperation. A good case in point is a collection by Blundo and Le Meur (2009) in which different organizational ways of collaboration are compared and evaluated.

If there is any science currently covering all three angles, it is sociology. There is a "sociology of markets" (Lie 1997; Fligstein and Dauter 2007), a "sociology of hierarchy" (Dumont 1966) and a "sociology of cooperation" (Infield 1971). However, the sociology's preoccupation is with groups, their dynamics, their interaction, their development. Therefore, a sociology of markets will never lead to a holistic understanding of all aspects in markets. It will not, for example, fully reveal the price effect of a monopolistic constellation. This also applies to the fields of hierarchies and cooperation. Sociology will only cover the (albeit interesting and relevant) aspects of group dynamics in the single realms of interaction. Another example of a social science which is not fixed in one corner of the triangle would be psychology. Psychology, like sociology, does not have a real focus on interactions as such. It focuses much more on individuals.

The archetypes being described rarely ever appear in their pure form. Cooperation, of course, play an important role both in companies and in markets. And there is hierarchy in markets as there are trades between employees in companies. No single

Fig. 1.2 The dynamics of interaction

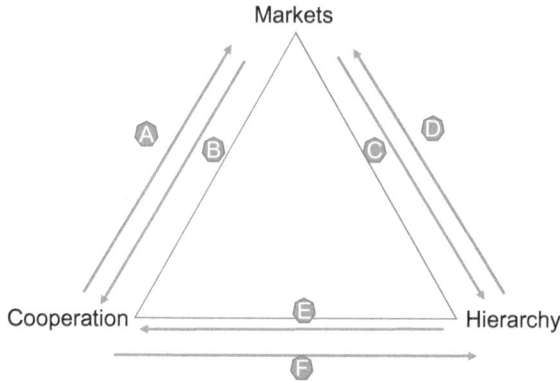

social science discipline as we know it, however, is prepared to understand these combinations. This starts with different toolboxes: The assumptions that psychologists make about individual behavior (driven by culture, parents etc.), for example, are totally inconsistent with what economists would assume (utility maximization). Nor would the single social sciences with their rather fixed assumptions regarding individual objectives, decision making processes or individual abilities be prepared to explain the dynamics in between the three presented modes of interaction. Such movements from one mode to another are to be described in the next section.

The following aims to show that many societal developments involve a shift from one form of interaction towards another. Using the arrows in Fig. 1.2, examples are provided for each of them.

A. Commercialization

Of all the arrows below, this is probably the one which receives the most attention from social scientists. Many scholars have argued that a number of realms in life are increasingly transferred from the cooperative sphere to the market, in both developed and developing countries. For the latter, the land market (Chabwela and Haller 2010) and the exchange of food (Lewis 1989) are examples being described by anthropologists. For developed countries, Wilkinson (2016) and Misra and Ghadai (2016), for example, focus on schools where an increasing part of content involves the purchase of certain items on the marketplace. Mann's (2008) examples of the societal shift from cooperation towards market-based exchanges include the area of problem-solving which has shifted markedly from family and friends towards psychologists. The most prominent example of a normative approach to this development comes from Sandel (2012). He suggests that public action should be taken to avoid further shifts from the sphere of cooperation towards economic exchange.

B. Community building

The move from markets towards cooperation can be understood in the spirit of communitarianism (Etzioni 1995) and is often driven by the desire to create strong and lasting human bonds and trust. That was the case in the movement to found Kibbutzim in the early 20th century (Russel et al. 2011), as it is in today's attempts at community-supported agriculture (CSA; Henderson and Van En 1999; Cone and Myhre 2000; Schnell 2007). Both are examples where participants attempted to link agricultural production with the creation of social capital, with bridging and bonding—in the case of the Kibbutzim among Jewish settlers, in the case of CSA between producers and consumers. The dynamics of this arrow is often driven by vision, impressively described by Conway (2016): "People, when they come together, can share understandings and manage their resources by enforcing norms and rules of their own design! The unconventional idea in many quarters was that people could cooperate beyond markets and states."

C. Mergers, Acquisitions and Integration

If there was one clear line in historic development of economic structures, it would be the enlargement of companies through mergers, acquisitions and other ways of integration, described as a "consistent and growing part of the business landscape" (Schweiger and Goulet 2000: 61). That this enlargement of companies is an implicit step from interaction through markets to the strengthening of hierarchies, is often an overlooked aspect. The most visible case of this aspect is probably vertical integration. Prominent examples in the literature are the auto industry and the agri-food chain. Particularly in the United States, car manufacturers have chosen to skip market relations to part-makers, instead integration the companies into the car-producing firm (Blois 2014; Langlois and Robertson 1989). In the realm of agribusiness, poultry production is a case in point for similar decisions. Vukina (2001) explains the tendency for meat processors to use contract farmers by factors like risk-sharing, innovation dissemination, demand for uniform quality and access to capital.

D. Outsourcing and "You Inc."

Management scholars have advocated that companies should focus on their core competencies and purchase all others on the marketplace (Quinn and Hilmer 1994), leading to the outsourcing of company parts to a self-reliant entity. Quinn (1999) describes that this process has begun in the realm of physical support like parts, componentry and hardware subsystems, and is increasingly shifting to intellectually-based resources which may be purchased more efficiently outside the company. In the extreme case, such processes lead to a situation where hierarchies are entirely replaced by market relations. Self-employment has become a societal trend and has spurred scientific discourses in two directions: in the Anglo-American world, advice has been given on how to turn this self-employment into economic success (Beckwith and Beckwith 2007). Under German conditions, the concept of the "Ich AG" has rather attracted concerns about precarious social conditions under which unprotected individuals may work (Sattelberger 2006; Keller and Seifert 2007).

E. Emancipation

If any long-term macro trend can be identified in the history of humankind, it may be summarized in this arrow. Emancipation has materialized in many different realms: the gender-based process of emancipation may the most visible and well-documented one (Markovic 1973), but the emancipation of the black race from slavery (Berlin 1985) and of homosexuals from discrimination (Steakley 1975) are other prominent examples. The tendency to substitute hierarchical structures by the spirit of even collaboration has also entered the business world. In particular, the Japanese concept of Kaizen (Singh and Singh 2009), while focusing on the steady improvement of process and product, has emphasized the "spirit of cooperation" (Brunet 2000: 3) which is necessary to unleash the full power of collective creativity. Styhre (2001) emphasizes the emancipation potential of Kaizen. Unfettering from external restraints is, in this case as in the other cases of emancipation, a crucial success factor.

F. New authoritarianism

In the realm of politics, many social scientists face a new breed of authoritarianism in a democratic setting, be it in Erdogan's Turkey (Taspinar 2014), Putin's Russia (Horvath 2011) or Orban's Hungary (Krasztev and van Til 2015). In an increasing number of states, voters are apparently willing to trade democratic gains as "necessary sacrifices on the altar of stability and growth", as McFaul and Stoner-Weiss (2008: 68) put it. Perceived or real, there may be shortcomings of strongly cooperative governance, and clear hierarchical structures may decrease a system's transaction costs. The view that "benevolent autocrats operating within a competitive institutional framework seem to provide the most effective combination of economic achievement and political restraints" (Gomberg 1966: 35) is also an accepted position among management scholars, so that the move from cooperative structures towards hierarchies is not restricted to the public sector, but also extends to companies.

Taken together, there seems to be a lot of "traffic" on the arrows in Fig. 1.2, and it can be argued that this traffic goes along all three sides of the triangle. This strongly dialectic situation offers many questions to be studied for socioeconomists. Apparently, the flows do not neutralize each other. Instead, they shape the look of a development which needs socioeconomists to be properly understood.

1.3 The Diversity Approach

Mainstream economists have developed a strong tradition of explaining the differences in growth rates between countries (e.g. Barro 2000). Increasingly, the focus of this analysis shifts towards the analysis of institutions: Particularly economists around Acemoglu and Robinson (2012) have emphasized the great importance of institutions for the development of nations. It has become obvious that institutions like low entry barriers in the financial sector, transparent standards in the educational system or flexible labor policies are able to contribute to economic prosperity.

One could term this worldview hegemonic or one-dimensional in the way that it implicitly assumes that there is one single optimal way of development for any nation you could imagine. Fukuyama (1992) has taken this idea to the extreme, suggesting that market capitalism in junction with liberal democracies are the ultimate and best form of governance, so that the end of history has ultimately be reached.

While the powerful movement in social sciences describing and explaining different varieties and eventually the resulting diversity of capitalism can be understood as a stand-alone approach, it is probably more fruitful to depict it as a countermovement to this impression. By taking this perspective, Peck and Theodore (2007: 732) describe how "the varieties approach represents a fecund interdisciplinary zone of engagement within the wider field of heterodox economic studies." After socialism had almost ceased to exist, it was neither useful nor necessary to depict capitalism as a homogeneous ideological block against collectivization and nationalization. Instead, an increasing number of social scientists started to describe the diversity of market-based systems. There was less normativity in this concept than in traditional growth models: Yes, some countries generated slightly less value added per capita than others, but maybe they fared better in other respect. Thus, the new, explicitly socioeconomic approach focused on describing different blossoms instead of measuring the most beautiful one.

Albert (1991) took a first step in that respect by describing the difference between a 'Rhinish' model typical of continental Europe and the classically neoliberalized path of Anglo-American countries, but a more thorough and formalized proposition was introduced by Amable (2003). He used cluster analysis to identify groups of countries with similar patterns concerning key socioeconomic variables such as labor organization, social policy and education. He found that countries with similar patterns were also often geographically adjacent, suggesting a strong cultural or at least geographical force behind the emergence of those systems.

The basic concept can also be traced back to the importance of institutions, observing their complementarity. A simple example would be the independence of central banks and inflation. The Central Bank's main concern usually is to keep inflation down, while governments are more interested in lush spending opportunities. That implies that the independence of the Central Bank will negatively correlate with a country's inflation rate.

It is easy to see that institutions play a key role in shaping the specific characteristics of capitalism in the country concerned, but less easy to understand why such different institutions would emerge. Hall and Thelen (2009: 7) emphasize the role of political equilibria: "Persistence of institutions depends not only on their aggregate welfare effects but also on other distributive benefits that they provide to the underlying sociopolitical coalitions." At least as important, however, is the role of culture in shaping such national systems. Bruff (2008) has explored this relationship in detail and, due to the many facets that the term culture can have, suggests referring explicitly to Gramsci's (1985) concept of common sense in society. Empirically, Schwartz (2007) is able to show that the observed differences in cultures can partly be traced back to people's value orientation. In attempting to explain the persistence of "varieties of capitalism", particularly in small countries, Elsner and Heinrich (2009) develop a game theory model indicating that homogeneous groups are formed within society in order to make cooperative behavior predictable.

It has been criticized that the "varieties of capitalism" approach is not very dynamic. There have been changes within capitalism, particularly of liberalizing economic restrictions around labor laws. If societies had created stable equilibria, this would have been very unlikely. Instead, it has proven useful to consider the possibility of shifts in a country: partly by linking the results from Public Choice Theory with the observations of systemic changes. Societal changes can often be explained by new coalitions between societal groups who press for such changes. Another explanatory approach are objective societal changes, such as the rising dominance of the service sector in the economy, or decreasing barriers of trade and communication.

1.4 Why Agriculture?

To the author's knowledge, this is the first book in which the emerging concept of socioeconomics is applied to an economic sector. Even for supporters of the idea of a bridging social science like socioeconomics, this requires two steps of justification: Why is it applied to one sector, and why to the smallest of the three (if looked from a worldwide perspective)?

Starting with the question on sectors in general, most economists and sociologists alike consider the distinction between the micro and macro level as sufficient. Indeed, what has been sketched so far in this chapter also either takes place on a micro level (interactions) and on a macro level (varieties). However, Elsner and Heinrich (2011) emphasize the importance of the 'meso' level which leads directly to the discourse on mesoeconomics, the analysis of economic sectors. It is probably fair to claim that, due to large-scale negligence in socioeconomic research, neither intrasectoral nor intersectoral dynamics are yet well understood, despite a few individual contributions on this issue (Ng 1986; Mann 2011; Dopfer 2012). The potential connection between the 'meso' and the 'varieties' approach becomes clearer when focusing on the important role that cultural specifications play in both theoretical frameworks. Few scholars would reject the widespread notion that there exists an industrial culture (Sadler and Thompson 2001; Mole 2007), a peasant culture (Viola 1999; Henningsen 2001) and perhaps even a service culture (Edvardsson and Enquist 2002; Skinner Beitelspacher et al. 2011) in all countries and, to some extent, also in regions.

It is therefore worthwhile to take into account the possibility that different cultures may lead to different interaction patterns and, ultimately, to different systems. This simple mechanism does not only apply when comparing geographic cultures (Chinese vs. American, for example), but also, when looking on single sectors with their specific cultural characteristics.

One of the arguments that could be made to support this claim is to dwell on the peculiarities of agriculture. The least that can be said is that it is the oldest of the three economic sectors and that the decrease of its relative importance is usually a good indicator for a nation's prosperous development. The sector's tradition also translates into organizational peculiarities: 80% of the world's food, in value terms, are produced by family farms (FAO 2014). 500 million of the world's 570 million farms are run by families. While similar things cannot be said about other sectors, this constellation has, of course, significant effects for the social setting in which production takes place. All production decisions have immediate effects on households, and all changes in the household like divorces or new children affect the production system with which a family farm operates.

Such interdependencies may also exist for the industry and the service sector, but they certainly are less obvious and more indirect than for crop and animal production. And that is where socioeconomic research can play its greatest strength: All families need a great degree of cooperation to function, and many of them also need some hierarchical structures. If these two interaction levels synthesize with the market on which most family farms sell their products, this provides a playing field where only holistic approaches such as socioeconomic research are able to see the whole picture.

One additional argument for using the primary sector as a case in point should be mentioned: In the farming sector, what is produced is not only the most basic of human needs, our food preferences also are extremely dependent on our social environment. "Food is culturally defined and acquires immense significance in all societies because it involves the human body." (Hamilakis 1999: 48). Again, there are cultural peculiarities (the domain of sociology) interfering with the sphere of markets (the domain of economists), so that another case for the application of socioeconomics could be made.

1.5 And Why Socioeconomics?

High quality textbooks on agricultural economics are available. So are very good textbooks on rural sociology. To justify the book at hand, we would need to show that a gap between these two breeds is to be bridged, and that nobody has done so yet.

In social sciences, many researchers look at causal linkages. Agricultural economists, for example, analyze how the wheat price in the EU is influenced by the wheat price on the world market. Rural sociologists check whether the self-understanding of second-home owners is dependent on the type of village they live in. But what if social factors are influenced by economic factors or vice versa? What if agricultural productivity is influenced by a growing number of female farmers? Or if the willingness to take over a farm shows to be dependent on income relations between sectors? Are such issues already addressed by the existing readings?

Agricultural marketeers could claim that this was the case. Arguably, they have been the first socioeconomists in the realm of agriculture, so as marketeers in general have been the first socioeconomists in social sciences. Because companies found out very early that social and economic factors were strongly interlinked. That men would buy more meat than women, but Muslim men wouldn't buy pork. Because gender and religion, among others, are important predictors of economic behavior.

Thus, marketing became, in a way, the first application of socioeconomic thinking. The whole concept of target groups rests on the fact that our economic preferences are strongly culturally shaped.

While this is an important point, this is certainly not all which can be said about the socioeconomics of agriculture. The primary sector entails many more choices which are culturally shaped. The way farmers produce, the support they get from the public, the way in which this support is organized—all these factors are results of social and economic factors in strong interdependencies. It is the objective of this book to show this.

References

Abell P (2003) On the prospects for a unified social science: economics and sociology. Socio-Econ Rev 1(1):1–26

Acemoglu D, Robinson JA (2012) Why nations fail: the origins of power, prosperity, and poverty. Crown Publishing Group, New York

Albert M (1991) Capitalisme contre capitalisme. Le Seuil, Paris

Amable B (2003) The diversity of modern capitalism. Oxford University Press, Oxford

Barro RJ (2000) Inequality and growth in a panel of countries. J Econ Growth 5(1):5–32

Beckwith H, Beckwith CC (2007) You Inc.—the art of selling yourself. Grand Central Publishing, New York

Berlin I (1985) The destruction of slavery: a documentary history of emancipation, 1861–1867. Cambridge University Press, Cambridge

Blois KJ (2014) Vertical quasi-integration. J Ind Econ 20(3):253–272

Blundo G, Le Meur P-Y (2009) The governance of daily life in Africa: ethnographic explorations of public and collective services. Brill, New York

Boyer R (2008) The quest for theoretical foundations: epistemology, methodology or ontology? Socio-Econ Rev 6(4):733–746

Bruff I (2008) Culture and consensus in European varieties of capitalism. Springer, Heidelberg

Brunet P (2000) Kaizen in Japan, vol 1, pp 1–10. From understanding to action (Ref. No. 2000/035), IEE Seminar, Kaizen, London

Chabwela HN, Haller T (2010) Governance issues, potentials and failures of participative collective action in the Kafue Flats Zambia. Int J Commons 4(2):621–642

Coase RH (1937) The nature of the firm. Economica 4(16):386–405

Coase R (1946) The marginal cost controversy. Economica 13:169–189

Coase R (1988) How should economists choose? In ideas, origins, and their consequences. American Enterprise Institute, Washington

Cone C, Myhre A (2000) Community-supported agriculture: A sustainable alternative to industrial agriculture? Hum Organ 59(2):187–197

Conway RT (2016) Ideas for change: making meaning out of economic and institutional diversity. http://www.countercurrents.org/2016/10/01/ideas-for-change-making-meaning-out-of-economic-and-institutional-diversity/ (23 Dec 2017)

Dopfer W (2012) The origins of meso economics. J Evol Econ 22(1):133–160

Dumont L (1966) Homo Hierarchicus – essai sur le système des castes. Gallimard, Paris

Edvardsson B, Enquist B (2002) 'The IKEA Saga': how service culture drives service strategy. Serv Ind J 22(4):153–186

Elsner W, Heinrich T (2009) A simple theory of 'meso'. On the co-evolution of institutions and platform size—with an application to varieties of capitalism and 'medium-sized' countries. J Socio-Econ 35(5):843–858

Elsner W, Heinrich T (2011) Coordination on "Meso" levels: on the co-evolution of institutions, networks and platform size. In: Mann S (ed) Sectors matter! Exploring mesoeconomics. Springer, Heidelberg

Etzioni A (1985) Toward socio-economics. Contem Sociol 14(2):178–179

Etzioni A (1986) Founding a new socioeconomics. Challenge 29(5):13–17

Etzioni A (1995) New communitarian thinking: persons, virtues, institutions, and communities. University Press of Virginia, Charlottesville

FAO (2014) Towards stronger family farms. FAO, Rome

Fligstein N, Dauter L (2007) The sociology of markets. Ann Rev Sociol 33(1):105–128

Friesl M, Sackmann SA, Kremser S (2011) Knowledge sharing in new organizational entities: the impact of hierarchy, organizational context, micro-politics and suspicion. Cross Cult Manag 18(1):71–86

Fukuyama F (1992) The end of history and the last man. Free Press, New York

Gide C (1905) Economie sociale. Sirey, Paris

Gomberg W (1966) The trouble with democratic management. Transaction 3(5):30–35

Gramsci A (1985) Selections from cultural writings. University of Minnesota Press, Minneapolis

Grey C (2004) Enterprise, management and politics. Int J Entrepreneurship and Innovation 5(1). http://journals.sagepub.com/toc/ieia/5/1 (Dec 19, 2016)

Hall PA, Thelen K (2009) Institutional change in varieties of capitalism. Socio-Econ Rev 7(1):7–34

Hamilakis Y (1999) Food technologies/technologies of the body: the social context of wine and oil production and consumption in Bronze Age Crete. World Archeology 31(1):38–54

Henderson E, Van En R (1999) Sharing the harvest: A guide to community supported agriculture. White River Junction: Chelsea Green

Henningsen P (2001) Peasant society and the perception of a moral economy—redistribution and risk aversion in traditional peasant culture. Scand J Hist 26(4):271–296

Hollingsworth JR, Müller KH (2008) Transforming socio-economics with a new epistemology. Socio-Econ Rev 6(2):395–426

Horvath R (2011) Putin's 'preventive counter-revolution': post-Soviet authoritarianism and the spectre of velvet revolution. Europe-Asia Studies 63(1):1–25

Ikeda K, Richey SE (2005) Japanese network capital: the impact of social networks on Japanese political participation. Polit Behav 27(3):239–260

Infield HF (1971) Utopia and experiment: essays in the sociology of cooperation. Kennikat Press, New York

Keller BH, Seifert H (2007) Atypische Beschäftigung - Flexibilisierung und soziale Risiken. Nomos, Baden-Baden

Krasztev P, van Til J (2015) The Hungarian patient. Central European University Press, Budapest

Langlois RN, Robertson PL (1989) Explaining vertical integration: lessons from the American Automobile Industry. J Econ Hist 49(2):361–375

Lewis MW (1989) Commercialization and community life: the geography of market exchange in a small-scale philippine society. Ann Assoc Am Geogr 79(3):390–410

Lie J (1997) Sociology of markets. Ann Rev Sociol 23(2):341–360

Lieshout RH (1995) Between anarchy and hierarchy: a theory of international politics and foreign policy. Edward Elgar, Cheltenham

Magee JC, Galinsky AD (2008) Social hierarchy: the self-reinforcing nature of power and status. Acad Manag Ann 2(1):351–398

Mann S (2008) From friendly turns towards trade—on the interplay between cooperation and markets. Int J Soc Econ 35(5/6):326–337

Mann S (ed) (2011) Sectors matter! Exploring mesoeconomics. Springer, Heidelberg

Markovic M (1973) Women's liberation and human emancipation. Philos Forum 5(1):145–153

McFaul M, Stoner-Weiss K (2008) The myth of the authoritarian model; How Putin's crackdown holds Russia back. Foreign Aff 87(1):68–84

Mikl-Horke G (2015) Traditionen, Problemstellungen und Konstitutionsprobleme der Sozioökonomie. In: Hedtke R (ed) Was ist und wozu Sozioökonomie? Springer, Heidelberg

Misra SN, Ghadai SK (2016) Quality education for all: a critique of draft education policy. Academicia 6(10):46–52

Mole T (2007) Byron's romantic celebrity: industrial culture and the hermeneutic of intimacy. Springer, Heidelberg

Müller KH (2015) Drei methodische Pfade für die Sozioökonomie im 21. Jahrhundert. In: Hedtke R (ed) Was ist und wozu Sozioökonomie? Springer, Heidelberg

Müller K (2016) Expanding Socio-economics in Four Dimensions. Forum Soc Econ (in print)

Ng YK (1986) Mesoeconomics. Wheatsheaf, Brighton

Ostrom E (1990) Governing the commons. Cambridge University Press, Cambridge

Pabst A (2012) Metaphysics: the creation of hierarchy. Eerdmans Publishing, Amsterdam

Peck J, Theodore N (2007) Variegated capitalism. Prog Hum Geogr 31(6):731–772

Quinn JB (1999) Strategic outsourcing: leveraging knowledge capabilities. Sloan Manag Rev 40(4):9–21

Quinn JB, Hilmer FG (1994) Strategic outsourcing. Sloan Manag Rev 35(4):43–48

Russel R, Hannemann R, Getz S (2011) The transformation of the kibbutzim. Isr Stud 16(2):109–126

Sandel M (2012) What money can't buy. Macmillan, New York

Sadler D, Thompson J (2001) In search of regional industrial culture: the role of labour organisations in old industrial regions. Antipode 33(4):660–686

Sattelberger T (2006) Die Irrungen und Wirrungen der Ich AG. In: Rump J, Sattelberger T, Fischer H (eds) Employability management. Springer, Heidelberg

Say L (1891) Economie sociale. Guillaumin et Cie, Paris

Schnell SM (2007) Food with a farmer's face: community-supported agriculture in the United States. Geogr Rev 97(4):550–564

Schwab S (1989) Coase defends coase: Why lawyers listen and economists do not. Mich Law Rev 87(12):1171–1189

Schwartz SH (2007) Cultural and individual value correlates of capitalism: a comparative analysis. Psychol Inq 18(1):52–57

Schweiger DM, Goulet F (2000) Integrating mergers and acquisitions: an international research review. Adv Mergers Acquisitions 1:61–91

Singh J, Singh H (2009) Kaizen philosophy: a review of literature. IUP J Oper Manag 8(2):51–72

Skinner Beitelspacher L, Richey RG, Reynolds KE (2011) Exploring a new perspective on service efficiency: service culture in retail organizations. J Serv Mark 25(3):215–228

Steakley JD (1975) The homosexual emancipation movement in Germany. Arno Press, New York

Styhre A (2001) Kaizen, ethics, and care of the operations: management after empowerment. J Manag Stud 38(6):795–810

Taspinar O (2014) The end of the Turkish model. Survival 56(2):49–64

Viola L (1999) Peasant Rebels under Stalin. Oxford University Press, Oxford

Vukina T (2001) Vertical integration and contracting in the U.S. poultry sector. J Food Distrib Res 32(2):29–38

Ward LF (1893) The psychological basis of social economics. Ann Am Acad Soc Sci 3:72–90

Wight M (2002) Power politics. A & C Black, New York

Wilkinson G (2016) Marketing in schools, commercialization and sustainability: policy disjunctures surrounding the commercialization of childhood and education for sustainable lifestyles in England. Educ Rev 68(1):56–70

Chapter 2
Agricultural Hierarchies

Hierarchies are a direct contradiction to equality: as soon as one person issues orders to another or is considered superior in any other way, these two persons can no longer be considered equal. Anarchists are the group who take this challenge most seriously and, in the extreme, even parenting can be considered as immoral, as it involves clearly hierarchical structures (Tremblay 2008).

On the other hand, hierarchies are often self-imposed, such as in our relationships with "celebrities", a keen focus of attention for many people. If you buy tabloid newspapers, watch TV shows or queue for tickets to see a particular group of people, this hierarchical structure between them and you cannot be so much of an evil. Media experts (Gorin and Dubied 2011) consider the rise and fall of celebrities to whom we "submit" ourselves in the public discourse as a way of negotiating social values.

In agriculture, hierarchies start in the farming family, but extend well into the relationships with associated businesses and the public administration. These three level fields for unfolding hierarchies will be covered in this chapter.

2.1 Public: The Agricultural Administration

All businesses need an institutional and legal framework in order to conduct their transactions securely. In many sectors in many countries, the degree of protection goes far beyond the provision of such a framework, but usually not as far as in agriculture. In countries such as Norway, Switzerland and Iceland, more than half of agricultural income is due to political intervention, and it is the public administration that has to provide the institutional framework for these interventions. By converting support from market intervention to direct payments in the 1990s in most Western countries, the public administration gained an even more central role. Whereas previously the administration had mainly administered certain purchases and sales of large quantities of commodities and imposed tariffs on the border, the introduction of direct payments necessitated exchanges with and controls of every single farm.

© The Author(s) 2018   15
S. Mann, *Socioeconomics of Agriculture*, SpringerBriefs in Economics,
https://doi.org/10.1007/978-3-319-74141-3_2

Looking at the agricultural administration, three levels can be distinguished:

– Farmers are mostly confronted with local administrations which are usually
  financed by regional or even local authorities. They receive farmers' payment
  applications, process them, check that all information in the application is correct,
  and hand out payments. Other divisions check applications for new farm buildings
  or for the refurbishment of old ones, or check whether all hygiene regulations are
  met in barns or farm salesrooms.
– At the other end of the spectrum, all national governments (and organisations such
  as the European Commission) also have their agricultural administrations. They
  work on political strategies which they transform into agricultural legislation, try
  to simplify or prevent agricultural trade, and represent their country's agriculture
  on international occasions.
– In many countries, particularly large ones, there are intermediate levels of the
  agricultural administration. Their task is to translate the legislative foundations
  into implementation or to design regionally specific programmes.

The following theoretical concepts usually apply to all three levels of the agricultural
administration, albeit often in different respects.

2.1.1 Weber's Iron Cage

In any writing on the rationale of the public administration, Max Weber should play
a prominent role, partly because the German sociologist (1864–1920) was among
the first to give the public administration a central position in sociological theo-
ries. Weber's general focus was on two concepts: one was rationalisation, which
he considered as the most general element in our historic development; the other
was domination, an apparently legitimate exercise of power. It is obvious that both
concepts are easily traceable in the apparatus of public administration. His famous
terming of the administration as an "iron cage" to describe both principles can be
found in his seminal work *The Protestant Ethic and the Spirit of Capitalism*:

> The Puritan wanted to work in a calling; we are forced to do so. For when asceticism was
> carried out of monastic cells into everyday life, and began to dominate worldly morality, it
> did its part in building the tremendous cosmos of the modern economic order. This order
> is now bound to the technical and economic conditions of machine production which today
> determine the lives of all the individuals who are born into this mechanism, not only those
> directly concerned with economic acquisition, with irresistible force. Perhaps it will so
> determine them until the last ton of fossilized coal is burnt. In Baxter's view the care for
> external goods should only lie on the shoulders of the 'saint like a light cloak, which can be
> thrown aside at any moment'. But fate decreed that the cloak should become an iron cage.
> (Weber 1905: 28)

This paragraph about the non-destructible institutional setting of administration,
like many other parts of Weber's writing, draws a rather bleak and pessimistic image

of the administration. While Weber describes the irresistible power of the administration in general, this certainly can also be applied to agriculture. In fact, the bureaucratic burden is particularly high in systems where farmers receive a lot of public support: in order to prevent abuse, checks are frequent and regulations are tight. But even in economies such as the Netherlands which traditionally provide a flexible framework for their entrepreneurs, "inefficient government bureaucracy" is among the top three factors considered as problematic for doing business among Dutch farmers (OECD 2015).

Being (or having been made) aware of the slack that an overwhelming bureaucracy creates for farmers, some governments have made an effort to reduce red tape. One example is the United Kingdom. A government commission (Department for Environment, Food and Rural Affairs 2014) reviewed the agricultural legislation, analysing which articles could either be deleted or improved. Out of 516 regulations, the commission recommended removing 156; it issued recommendations to inspect farms less frequently and to dispense six monthly reports on the numbers of mosquitoes imported for research purposes.

Such attempts fit in well with Weber's concept of formal rationality. In this, he described the attempt to establish organisational forms which are as resource-efficient as possible. In Weber's view, bureaucratic administration was the primary way in which rational-legal authority has developed in formal organisations. There was such a broad general acceptance in society that the administration would not need to defend its legitimacy; it could fully focus on its task to find logical solutions for the organisation of public life. For the agricultural administration, both forces can be considered in this context: the desire to regulate all cases by issuing additional laws and orders, but also the desire to simplify public life by cutting red tape, rationalising the interactions between farmers and the public administration.

2.1.2   Niskanen's Bureaucrat

While it is possible to understand Weber's sceptical view of bureaucracy without a large body of prior knowledge, this is not the case for Niskanen's (likewise sceptical) perspective. William A. Niskanen built strongly on Public Choice Theory, which should be introduced briefly at this point.

While traditional economic theory focused on the consumer's wish to maximise utility, economists would often also (mostly implicitly) assume that policy-makers would steer this process in the best possible direction. Early public choice theorists found this unsatisfactory, and inconsistent with what they observed in the political arena. What would happen, they asked, if the assumption about individual utility-maximisation were extended to the breed of policy-makers? Models were developed that showed, for example, how political parties, in attempting to maximise their votes, would target and compete for the median voter. Or how interest groups maximised the numbers of their members or their political influence.

For Niskanen, who started his career in the US Federal administration before going into academia, it was only a small step to extend this concept to the public administration. If consumers wanted to maximise consumables and political parties wanted to maximise votes, what was the objective function of someone in the administration?

His answer in his 1971 book *Bureaucracy and Representative Government* was: the budget. If a person in the administration could choose between a large and a small budget, she would typically choose the large one for two reasons: one, (potential) recipients of the money would try to please her (not necessarily by outright corruption, but at least by an increased degree of attention). Two, both her wage and power would be strongly dependent on the level of her budget. It is true that most salaries in the civil service are standardised by the public wage system, but someone with a million euro budget will probably be put into a higher wage class than a colleague with a budget of 10,000 euros. Niskanen's book has therefore been cited in more than 8000 scientific publications, and his conceptual approach has never really been challenged.

How does this translate into the realm of agriculture? Let us use the German Agency for Non-food Uses (Fachagentur Nachwachsende Rohstoffe e.V.—FNR) as a case in point. This agency was founded in 1993 to distribute money for research into the non-food uses of agricultural materials on behalf of the German Ministry of Agriculture, ranging from new ways to convert wood into heat to ways to develop lubricants from vegetable oil. In 1994, it had 20 employees to administer a budget of 50 million German marks (26 million euros). In 2017, the same agency had 93 employees and a budget of 61 million euros.

It is likely that Weber and Niskanen would have different perspectives on this success story. Weber would probably stress the increased relevance of non-food uses from agriculture. Would not the increased need for a carbon-neutral economy justify any budget rise for the use of natural material as the most rational decision? Maybe, Niskanen might answer, but why would increasing the budget by a factor of 2.3 necessitate increasing the number of staff by a factor of 4.7?

Whatever good reasons could be found to justify the larger staff of this agency (and, of course, of many other government organisations in many countries), it is likely that Niskanen's approach would provide a reasonable contribution to explain the development. Public service organisations certainly have dynamics of their own, and in any scientific appraisal of the administration this should be one of the aspects to consider.

2.1.3 Principal-Agent Issues

With Niskanen's concept of the budget-maximising bureaucrat, the epistemological potential of 'maximising utility' in and through organisations had certainly not come to an end. Only a few years later, the economists Michael Jensen and William H. Meckling brought this paradigm into the middle of hierarchical structures by developing the principal-agent model (Jensen and Meckling 1976). This model takes into account that, within a hierarchy, the promising strategy to maximise utility of the superior person (the principal) will be different from that of the subordinate person (the agent).

The foundation of these different positions (and resulting different optimum strategies) is information asymmetries. Imagine a hierarchy in which the agent is obliged to accomplish a certain task for the principal. And consider that it is usual for the principal to have less information on this task: how long it takes, how difficult it is to complete, and what the properly completed task should look like. Usually, the principal would supervise several agents who are contracted to do tasks that the principal himself may never have done.

Such asymmetries cause moral hazards. The agent will have the opportunity to exaggerate the effort he is putting into his duties, causing excessive financial demands. Or, similarly, he reduces the time and effort he invests and delivers an inferior output, claiming it would be the best possible solution. Of course, all of this only works when the agent has no significant share in the outcome.

The world is full of principal-agent problems; they occur in business as well as in politics. But the agricultural administration certainly has its fair share of them too, particularly (again) in systems where the state interferes more strongly in farmers' activities. They occur within administrative units, between different levels of the administration and between the administration and farmers. The following transcribed text sequence is chosen to illustrate this latter point.

The sequence is taken from an interview with a regional farm controller in Switzerland, a country with extremely strong (supportive) interference in farming. The interview had reached the issue of shortcomings in terms of controllability of direct payment programmes, specifically the payments for grassland-based milk and meat production (GMM). It quickly shifts to the Resource Efficiency Payments (REP) where farmers receive additional money if they register for no-tillage, use drop hoses or apply pesticides with a protective technology. Interviewer I2 is a part-time farmer himself and changes role over the course of the sequence.

C: We can discuss till doomsday. But this is not only the case for GMM. REP, REP, REP is actually much worse. No-tillage, that is actually much worse.

I1: Are you going outside and look at the soils?

C: Yes, yes, what do I see, what do I see now? Do I see whether the wheat, whether the wheat has been grown with mulch-till?

I2: A heap of rubbish. You have to register, you prepare, you say I am doing mulch-till, mulch-till, mulch-till. I have also this year, I registered for mulch-till, I grew leek for the first time, right? And now the leek has come later and later, until June 30 you had to register, so I said, what do I do, the leek is not inside yet, but I have registered mulch-till. If now the controller arrives, no matter what you say, he just has to believe me, full stop, doesn't he?

C: Yes, and particularly, I generally have to believe if it is mulch-till. I do not see what has been done with the seed. So it is exactly the same at this point. And therefore

I say, it is a, I think one should, one should control things which can be controlled, not only just believe, shouldn't you? But it is difficult.

The controller's intention seems to be to put the points being raised into a broader context. Resource Efficiency Payments are chosen to produce (probably) the worst possible example. The many repetitions catch the eye, a pattern that has been correlated with oral and unplanned discourse and with self-reference. The statement that REP is worse in terms of controllability than GMM is now specified with regard to the three bricks which make up REP. It is no-tillage rather than drop hoses or pesticide application which causes most of the pain.

Interviewer 1 takes up the emotional drive by switching into present tense, even though the controller will hardly go outside during the interview. In terms of content, however, Interviewer 1 ignores both the normative and emotional content of what has been said, restricting his focus to the control's organisation. He suggests what a control could look like, and it is not entirely clear what purpose the controller's "yes" actually serves. In any case, the controller subscribes to the image of him going outside now, applying the present tense himself. The core problem, the missing possibility to observe no-tillage on the farm, is transformed into a rhetorical question. It seems obvious enough that it is impossible to control mulch-till on the field, so it is not even necessary to put this into a statement.

This is where Interviewer 2's story comes in, starting with an only-normative statement with which he affirms the controller's attitude. The story then circles around the incompatibilities between the phasing of the application for REP and the production phases for leek. His point is that controllers would have no factual evidence to check the compliance with no-tillage.

When the interviewee speaks again, he affirms what has been said, even though he wants to make his own case. Although he introduces this with a "particularly", the opposite would be more correct. This is not a special case in the general remark by Interviewer 2; rather, Interviewer 2's story is a special case in the general concern of the controller. He now answers his rhetorical question, apparently doubting whether his underlying point was understood before. He then needs a few attempts to draw his general conclusion. The "I say" denotes subjectivity, whereas the "it is a" signals a high degree of objectivity. He then steps back from this in order to finally choose "I think", carefully enough for the fact that his sentence that "one should control what can be controlled" should go largely undisputed.

Taken together, it is clear that the controller takes the role of the principal, complaining about the impossibility of controlling the farmer (the agent). Although this is, of course, an extreme example of information asymmetries and indicates that the agri-environmental programme was not well designed, asymmetries and resulting moral hazard issues are omnipresent and often mentioned in the agricultural press. Principal-agent theory seems to be a helpful tool to understand the dynamics of the agricultural administration.

2.1.4 New Public Management (NPM)

While Max Weber had emphasised the peculiarities of public administration, the NPM movement had the opposite motivation, connected with a strongly normative message. Public administrations would not have to be so distinct from profit-oriented enterprises, and the closer they became, the better. This was the core message of British Prime Minister Margaret Thatcher in the 1980s who wanted to make local administrations in particular more efficient. It was also subsequently the message of scholars such as Pollitt (1993) who showed ways of transferring management principles from private companies to public agencies.

Possible pathways to make public administration more efficient could be identified in many areas. One of them was the setting of incentives, where it was suggested that public employees be paid according to performance rather than given fixed wages; another was accountability. By applying performance standards and output controls, it should be possible first to obtain clarity about the different cost levels of an administration and finally to cut costs by discovering and removing inefficiencies.

Although it originated in Britain, NPM spread quickly, first within the English-speaking world, but soon to developing countries and continental Europe. In addition, international organisations such as the OECD have established working groups on NPM. The idea that strategies could be established to increase the efficiency of the "iron cage" has broadly attracted strong levels of support.

Not all attempts at applying NPM have been equally successful. Hubbard (1995), for example, shows that performance contracting for public services for agriculture had been attempted, but was too resource-demanding for administrative bodies so that it ultimately failed.

Finally, NPM is not much more than a rather general idea—administrations should work more like businesses—which can be translated into practice in very different ways. The agricultural administration has not been at the forefront of these attempts. Nevertheless, ideas to drive efficiency in the public sector have also influenced research concerning the agricultural administration, as we will see in the next section.

2.1.5 Production Economics

Traditional production economics goes back to Frederick Winslow Taylor (1856–1915) who was concerned with optimisation of production processes, leading to maximisation of outputs with given costs or minimisation of costs with constant outputs. It was not a far step from New Public Management to apply this principle to the public administration. While production in its strict sense would not happen in the administration, it had already been recognised that the costs of administering issues, a part of what economists called transaction costs, were no less important. Their systematic analysis was first carried out in the tax sector. Several scholars

(Sandfort et al. 1989; Grüske 1991; Allers 1994; Raab 1995) focused on the costs required to collect one euro in taxes, finding that some tax categories caused far higher transaction costs than others.

Generally, there are two ways of estimating the costs of political programmes. After identifying all of the organisations involved, the direct method is to ask accountants in each organisation for their cost estimates. The indirect method is to make these estimates yourself by analysing organisational charts and the organisations' budgets.

Again, only a small step was needed to transfer these questions and methods to the realm of agriculture and to develop them further. Like tax collection, all political instruments used to support farmers would necessarily entail a certain level of transaction costs. In a first wave of research inspired by the methodology of production economics, these costs were to be measured and explained. Among the findings of this phase were the following results:

– When support programmes were cut, this did not necessarily cause lower administrative costs. During the time when EU export subsidies were radically reduced, the costs for their administration and control were still on the rise, due to tighter control activities (Mann 2002).
– The nature of support programmes would strongly influence the level of transaction costs. Specific programmes to which farmers rarely subscribed caused higher transaction costs (on the administration's side, but also on the farmers' side) than general and broad payments (Rørstad et al. 2007). This finding was occasionally used to defend general market support as compared to more target-specific measures.
– Another factor influencing the level of administrative costs was the administration's organisational structure. Multiple levels of hierarchy between the policy-making unit and the unit handing out payments would increase transaction costs considerably (Mann 2001).

In the long run, however, it could not be sufficient simply to measure and explain different cost levels of different political instruments. Yes, some policies would cause higher transaction costs than others, but what would that imply? Might these "expensive" policies not be much more effective than policies with low levels of transaction costs?

This research is still in its absolute infancy, mostly due to the difficulties of defining a policy's real impact in terms of success. Fährmann and Grajewski (2013) made a first attempt in this direction. They asked experts to estimate the impact of a series of different rural developments. And they measured the administrative costs of the same policies, allowing them to estimate correlations. Unfortunately, their results were somewhat ambiguous. Over the entire sample, they could not find any meaningful and significant relationship between the estimated impact and the cost level. Only in one of their regions (Hesse in Germany) was it possible to say that programmes with a high impact level caused higher administrative costs than programmes with a low impact level.

2.2 Commercial: Power in the Chain

Everybody knows that the administration works hierarchically and that our own relationship with the administration is based on a hierarchical structure. For economic transactions, the situation is different. When economists look at markets, they like to think of individually utility-maximising agents on an equal footing, at least in the classical models. Only recently have economists become increasingly open to the notion that power structures also exist in markets, and have devoted attention to the nature of such structures. The following section will highlight some examples of asymmetric market structures of relevance to agriculture, together with their causes and consequences.

2.2.1 Getting Started: Price Transmission

Based on the conventional microeconomic model world, the rational actor should transmit 100% of price changes. Imagine a dairy which buys milk for 40 cents per litre and bottles it, selling it for 50 cents per litre. As soon as the farmgate milk price drops to 30 cents per litre, the price per bottle will immediately drop to 40 cents, if a few simplifying assumptions (such as inelastic demand) apply.

Even without power asymmetries, this does not depict real-world conditions. In fact, many factors in the economy put price transmission rates well below 100%—and produce considerable time lags. One of them is distance. There is always a time lag between the first and the second price change, but Mengel and von Cramon-Taubadel (2014) show that 1000 km of distance within a country decrease the speed of price transmission by 6–20%.

But it is obvious that power relations may also have a major impact on the speed (and occurrence) of price transmission. Once producer prices for milk have fallen, our invented dairy will probably attempt to keep its selling price at 50 cents for as long as possible. A factor that could force the dairy to pass the farm price decrease on to consumers could be, for example, that a competing dairy has done so. Indeed, economists have shown theoretically (Weldegebriel 2004) and empirically (Muslim 2011) that oligopolistic structures in the chain slow down or even prevent full price transmission.

Vavra and Goodwin (2005) provide an example of a lagged and incomplete price transmission in Fig. 2.1 which can be applied to our milk and dairy example: the dairy in this illustration immediately starts to pass on the price increase, but cannot fully pass on the shock experienced at the farmgate level. Only gradually, with decreasing speed and ultimately incompletely, is the price increase passed on to consumers.

This example suggests a weak power position of the dairy as compared to the farms. The opposite would be much more typical. Agricultural economists largely agree that farming, particularly in small-structured systems, suffers from a structural disadvantage. Many farmers deal with a few fertiliser, pesticide and tractor producers

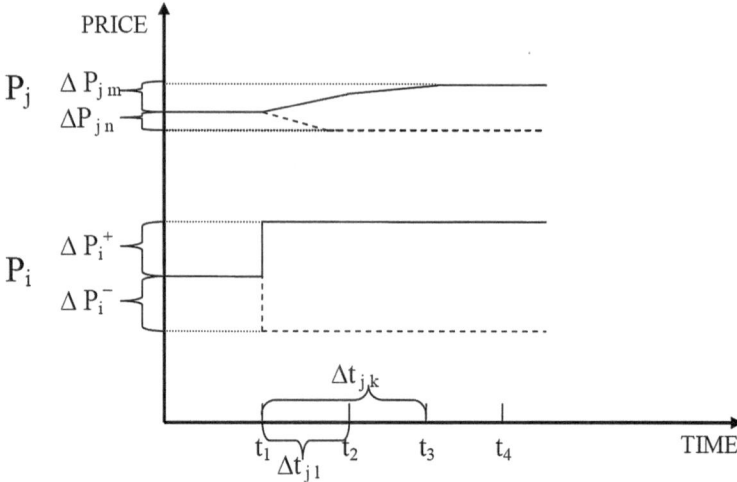

Fig. 2.1 Illustration of an asymmetrical price transmission (Vavra and Goodwin 2005)

on the one hand, and with a few slaughterhouses, mills and dairies on the other. Empirical evidence from all over the world indicates that lags and limitations in price transmission rarely work to the advantage of farmers (e.g. Balisacan et al. 2010; Gembreselassie 2012), nor to the advantage of the poorest groups in the world's population which are in certain times hit hard by spiking world prices (e.g. Cudjoe et al. 2010).

This section, like the following sections, shows a different case of hierarchy than the type of hierarchy encountered around the agricultural administration. Between the administration and the farmer there is a legal or **explicit hierarchy**. If a public controller comes onto the farm, he may be smaller, weaker and less smart than the farmer. Nevertheless, he is in a superior hierarchical position, ultimately due to the state's monopoly on power. This is different when the farmer and the dairy negotiate the milk price. They may be legally on equal terms. However, the dairy manager is in a much better position to reject the farmer's price suggestion than vice versa. This should be termed an **implicit hierarchy**.

2.2.2   The Special Case of Land Grabbing

In most cases of major societal dispute such as GMOs or abortion, the opposing parties at least agree largely on a common terminology. In relation to large-scale land acquisition or land grabbing, this is not the case, and it is surprising that two discourses about the same subject can be so decoupled from each other.

Many economists generally support investment from richer nations in poorer countries, termed Foreign Direct Investment (FDI). 'The contribution of FDI as key participant in economic growth has been widely acknowledged,' wrote Mughal and Akram (2011). FDI is considered a key instrument for the modernisation and diversification of lagging economies. If wealthy and innovative businesses cannot fix the problem of poor environments, who should?

The last 20 years have seen unprecedented growth in FDI involving land. Mostly, investors have been major enterprises from industrialised countries, such as the Daewoo Logistics Corporation or the German Neumann Kaffee Group. In some instances, governments act themselves. The Chinese administration in particular has become active and, for example, has leased 2.8 million hectares in the Democratic Republic of Congo in order to manage the world's largest oil palm plantation (Baxter 2010).

As soon as investments focus on land, FDIs change their name. They are called large-scale land acquisitions (LSLA) by their supporters, who can be found more in the realm of politics than among academics. African governments in particular often strongly encourage foreign investors to invest in land and produce cash crops on a large scale (Abbink 2011).

The group of opponents to what they call "land grabbing" is probably more numerous, or at least more vocal, than the supporters. While many academics have criticised the ways in which Northern enterprises work on Southern land, the centres of resistance are NGOs such as Oxfam or Bread for All, for which the struggle against land grabbing is a worthy source of donations. 'The global rush for land is leaving people hungry,' they argue (Oxfam 2017), promising to stand for the rights of expropriated smallholders in the affected countries.

Well, what is the problem? How could FDI, as soon as land was involved, turn into such a contested issue? Tools to develop an answer to this question certainly have much more to do with the asymmetries of hierarchies than with market transactions. In most cases, we have traditional smallholder systems on one side, often dominated by slash-and-burn agriculture with very low productivity. On the other side, modern agronomic systems with the optimised use of contemporary farm technology are to be implemented, making ample use of scarce water resources and pesticides. The managers in charge of the Northern companies and local smallholders are an odd group of competitors for farmland.

In fact, one could argue that many of the "land grabbing" cases qualify as the most radical land use changes in history. Over the course of human development, land use changes have tended to come gradually and slowly, be it the change from three-field crop rotation to continuous arable farming, or from pastoralism to more intensive grassland management. The most revolutionary example of land use change was probably the collectivisation of Soviet land under Stalin, but even there the modes of production changed only slowly, despite a radically new mode of ownership. For most LSLA projects, however, everything changes, mostly from one year to another: ownership goes into the hands of a major enterprise, the portfolio changes from diversified to specialised and often from staple crops to cash crops, and the degree of intensity multiplies.

This is another classic example of an implicit hierarchy. It becomes clear that the grave power asymmetries between the actors involved are a large part of the problem.

If a peasant in Sierra Leone and a Chinese company with a massive turnover compete for land, this is not a real competition. Even acknowledging the positive effects of FDI also in agriculture—particularly boosted productivity—it is probably a good thing that the public closely watches the conditions under which the land transfer is taking place.

The devil lies in the detail. The Coca-Cola Company, for example, attracted attention by announcing that it would not accept sugar deliveries from land being taken from smallholders. 'The Coca-Cola Company believes that land grabbing is unacceptable' (Tran 2013). However, reports issued to demonstrate that all goods used for Coke production are unrelated to land grabbing raised criticism from the NGO side as being 'too superficial' (Dawson 2015).

That leads to the necessity to establish broadly accepted and credible institutions which define acceptable and unacceptable ways of agricultural production, for FDI and beyond. This aspect will be taken up later (Sect. 4.3).

A final anecdote may clarify the relationship between land grabbing and hierarchies. The Swiss enterprise Addax Bioenergy invested in 30,000 ha of sugar cane and an ethanol plant in a poor and remote area of Sierra Leone. A teacher in one of the affected villages was asked about his position towards this investment. He conceded many positive effects, but then complained about the pitiful state of a pedestrian bridge leading to his village: 'Addax should definitely do something about it.'

Usually, it is the local administration which is responsible for the maintenance of local infrastructure. However, there are areas in which these institutions are extremely weak or non-existent. A billion-dollar company from abroad can, under such circumstances, quickly come into the role which, under normal circumstances, would be taken by public authorities. This implies that de facto hierarchies between locals and this company become extremely similar to the hierarchy between locals and a functioning administration.

2.2.3 Vertical Integration

Horst Bühler, a meat packaging enterprise in Southern Germany, is looking for organic turkeys. But it's not done to call them to sell a few hundred on the spot. Horst Bühler only buys turkeys on long-term contracts. These contracts run for five years if you are building a house for the birds, and three years otherwise. And they contain more details than one would expect. This starts with the birds themselves, which have to be bought from one particular breeder. And it continues with the feedstuffs. Farmers have to buy five ingredients sold by two mills and to feed them in a pre-defined formula. Only just before delivery are they allowed to add their own cereals to the feedstuff that comes at a uniform price negotiated between Bühler and the mills. After 21 weeks, the turkeys are collected by Bühler.

By now, it will be clear why this paragraph has been placed in the "hierarchy" section. The farmers delivering to Bühler receive good prices, they have long-term security for their product marketing and they produce good quality. But they have to

Fig. 2.2 A model of vertical
integration in agriculture

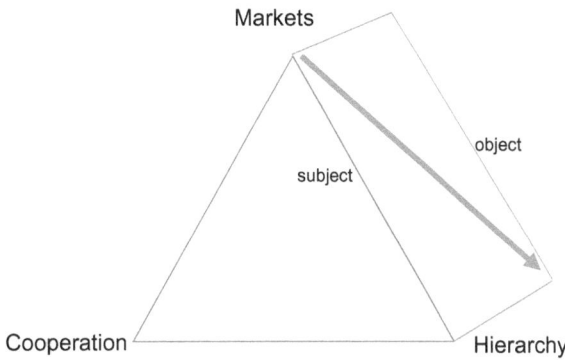

submit totally to regulations developed by others. In the life of a turkey producer in this setting, there is not much entrepreneurial freedom left. Examining the case of German pork production, Schulze and Spiller (2006) observe that 'vertical integration represents hierarchical governance mechanisms'.

However, the more that farmers become part of business structures where they largely follow production rules—including which breeds to rear and which feedstuffs to use—the less they are self-reliant subjects, and the more they become objects driven by their environment, a development illustrated by Finan (2007) and Tuong (2009). Therefore, it may be worthwhile to simultaneously analyse the interdependencies between the two movements: one from markets towards hierarchy, and one from subject to object.

Figure 2.2 extends the theoretical framework of Chap. 1 by adding the dimension of subject-object perceptions. Economists usually prefer to envisage individuals as subjects with a clear preference structure which they aim to cover. Sociologists tend much more to emphasise the position of individuals as results of their culture and environment. Socioeconomics could be a discipline suitable for the combination of both perspectives. And the vector in Fig. 2.2 uses the case of vertical integration to demonstrate the changing position of farmers in the business world.

It would now be useful to draw our attention to the advantages of vertical integration, taking into account that both Horst Bühler and its poultry producers entered voluntarily into this arrangement. Fan et al. (2014) mention some central arguments, including:

– Information asymmetries between participants
– Agency problems inducing opportunistic behaviour
– Dysfunctional institutions and market forces preventing contractual arrangements.

It is not always straightforward to apply such general findings of institutional economists to the context of agriculture. Sometimes, arguments are applied somewhat arbitrarily. When Klein et al. (1978) write about contractual arrangements concerning farmland, they call agricultural land 'a highly specific asset' (p. 320), while Allen and Lueck (2003), on the same issue, argue that 'Landowners bring just one asset to the exchange: land—an asset that typically is not specific to the exchange' (p. 36). In both cases, the classification was crucial to explain their respective empirical results, i.e. to bring them into accordance with institutional economic teachings. The fact that they used contradicting premises, however, shows that these teachings are probably yet not as consolidated as they should be.

A study by Dries and Swinnen (2004) links the issue of vertical integration with that of FDI, using the case of Polish dairies. After the fall of socialism in Poland, industries in the country were either bought by Western enterprises or kept by local competitors which, however, attempted to adapt to technologies and organisational structures introduced by their new competitors.

Poland's agriculture arguably has the smallest structures of all major European countries. And modern dairy industries are not too helpful as long as smallholders cannot produce milk of sufficient quality. Therefore, many dairy companies offered assistance programmes to delivering farmers. And once foreign companies started offering feed supply programmes, trade credit and investment assistance programmes, domestic companies soon followed. The two Belgian researchers were able to show that both farmers and dairy companies were more successful if such elements of vertical integration were used to make Polish milk production more competitive.

2.3   Private: Powerful Families

Family farming is the dominant form of today's agriculture. One can argue about what exactly can still be counted as a family farm, but it cannot be argued that the vast majority of today's 570 million farms are family farms. Lowder et al. (2016) estimate that 75% of the land is farmed by family farms. And behind each family farm there is, of course, a farming family.

Two justifications can be found for the choice to group families in the "hierarchy" section, although family life also contains elements of cooperation and markets (as will be shown soon). The first of them is historical.

The first publications that today could be termed under the sociology of families appeared well before the discipline of sociology was consciously established. Nave-Herz (2013) beautifully describes these first attempts, being motivated by the sorrow for preserving this social institution, and depicting the concept of a clear hierarchical structure as an ideal solution.

The dawn of the 20th century saw the rise of emancipatory ideas regarding gender. The call for cooperative management of marriages became increasingly impossible to ignore. On the inter-generational level, the late 1960s then saw the rise of anti-authoritarian educational approaches. The idea of the entire family as an organisation of cooperation was born.

In 1981, the economist Gary S. Becker opened another perspective. Could family decisions not be portrayed as market decisions? One important example given by Becker was the marriage market; another was the demands of children. However, the explanatory power of pure individual utility-maximisation in family decisions remains limited: not only did Becker face criticism for his emphasis on economic reasoning (e.g. Granovetter 1985), he decided himself to devote a chapter of his *Treatise* to altruism in the family. For an open mind such as Becker, it was easy to see that individual utility-maximisation on the market could never tell the full story of founding (and living in) a family.

Not only are hierarchies the oldest interaction mechanism under which the family was viewed: a stronger argument to subsume the family in the "hierarchy" section is the fact that the agricultural family is always a family defining a hierarchical business.

Family businesses are a unique and rich research field because they link the complex structures of businesses with the complex structure of a family. The boundaries can rarely be defined: 'Consider the case of a Canadian prairie farm woman who drives 40 min each way, across the large expanse of an ever-growing and ever more industrialized family farm, to deliver meals to the field for farm workers. Is this domestic work or farm work? [...] On family farms, there is often little separation between "waged" and "unwaged" labor.' (Fletcher and Kubik 2017: 3).

While a family farm may be incredibly multifaceted, it is based on two different dimensions: one is the relationship between the farming couple (and although "queer" farm couples are not unheard of, it is still a fairly good proxy to talk of inter-gender relations), and the other is the relationship between generations.

2.3.1 Inter-gender Relationships

Few terms are valued so differently between the sexes as "traditional family farming". For many men (e.g. Jager 2004), traditional family farming is a precious heritage that should be maintained. For many women (e.g. Bektasoglu 2012), traditional family farming is an oppressive and unjust division of rights and labour between men and women.

One does not have to be a gender expert to understand these differences. Even a superficial look at the literature is sufficient. In a study about Sub-Saharan Africa, Puppin-Lerch (2007: 25) finds that:

> Women played an important role in agricultural production and the following market trade. Women ploughed, planted, weeded, and harvested the field; transported their products from the field to the village; and sometimes marketed the farm products. Nevertheless, although farmers were mostly women, men were the owners of the land. They also owned the cattle, a major source of wealth and power.

This short paragraph shows two important characteristics for traditional inter-gender relations in farming. One is clear power asymmetries: men are usually the owners of most assets, even if most of the work is done by women. The impacts of many millennia of male domination are today, in a time that strives for fair inter-gender relations, most visible in farming as the most traditional economic sector. In fact, the FAO (2013) estimates that the amount of land owned by women is still below 10%, as is the volume of extension services from which women benefit.

The second pattern is the rather fixed and inflexible division of work between the genders. Such a fixed and defined distribution does of course save transaction costs. However, the shape taken by this division of labour varies greatly between cultures:

– In many developed countries, women tend to work off-farm and men on-farm. This is, for example, what Muenstermann (2010) reports from Australia.
– Higgins' and Fenrich's (2011) observation that men tend to be responsible for cash crops while women produce staple crops also applies for a number of other Southern hemisphere countries.
– In many pastoralist systems, women are responsible for milking the cows, even if men are in charge of milking other animals (Niamir-Fuller 1994).

At this point, it seems worthwhile to leave the practice of family farming briefly in order to raise a sociological concept, namely that of social construction. Berger and Luckmann (1966) suggest that reality is always a constructed reality. Certain persons, certain characteristics and certain objects are meant to play a certain role. If we see a person in a green uniform and with a gun, we will quickly construe this person in our minds as a soldier. Our ideas about his role in society will go far beyond his green clothing and his weapon. Berger and Luckmann, in such cases, would opt for deconstruction. Especially if the roles we were ascribing would cause unhappiness, it is beneficial to broaden the possibilities of roles and behaviour in society.

In a similar way to the soldier example, most of us ascribe roles to male peasants or farmwomen. Williams (1989) was among the first to suggest that gender should also be deconstructed. In rural societies there were usually fixed patterns about the role to be taken by women. These patterns had the advantage of lowering transaction costs. If it was clear that women would do the milking and men the ploughing, then this division would not have to be negotiated in evolving partnerships. However, the shortcomings of such patterns are also obvious: they are usually not well adapted to individual preferences. There may be numerous cases in which women milk and men plough, although both would be happier if men were to milk and women were to plough.

From this perspective, Rossier (2004) portrays seven Swiss farm couples. She finds cases where man and woman are stuck in traditional roles which make them unhappy. But she also finds cases where couples manage, going beyond traditional roles, to find divisions of labour which are well adapted to individual needs and interests. Which do not necessarily imply that both persons are involved in farming at all.

However, this "solution" of deconstruction may be unduly idealistic. If men have the power in the system, what should motivate them to let it go? Thus, some rural sociologists (e.g. Allen 2007) argue that the process of enabling farmwomen and appreciating their work will necessarily entail conflict.

Is agriculture, after all, becoming more male or more female over time? This discussion was initiated in 1978 by a Romanian man (Cernea 1978) who argued that women would play an ever-increasing role in the farming sector and that a "feminisation of agriculture" was underway. In the decades to follow, this discussion was continued with some enthusiasm. While it has been shown that women farm differently from men (farms led by women are usually smaller and more often organic), however, it has not been effectively shown that such feminisation would occur on the global farming scale. Tendencies go in both directions.

2.3.2 Inter-generation Relationships

At the beginning of this chapter, the relationships between parents and young children were used as the prototype for hierarchical relationships. But to what extent do these parent-child relationships impact the farming sector as a whole?

It is usual for the manager of a family farm to retire and sell or rent his farm to somebody from the neighbourhood or a different part of the world. Such inter-family successions are fairly frequent in the New World compared to more traditional European systems, and also in certain segments such as periurban horticultural businesses (Bertoni and Cavicchioli 2016). However, on a global scale, it is much more usual for farms to be handed over within the family, within the very hierarchical relationships between parent and child.

Economists do not usually deal with parent-child relationships. Nevertheless, the 1990s saw an emerging interest among agricultural economists in the process of structural change. In developed countries it was clear that the number of farms had been declining for decades. Germany, for example, has seen its number of farms shrink from more than one million in 1950 to less than 300,000 in 2016. The turn of the century brought a large number of publications that econometrically explained under which circumstances this process of decline would be speeded up, and under which circumstances the structure would remain stable. This wave of research (for an overview see Mann 2003) generated the following main results:

– The older the farm manager, the more likely it is that a farm will be abandoned.
– The larger a farm, the less likely it is that a farm will be abandoned.
– Part-time farms are more likely to be given up than full-time farms.
– Direct payments decrease the probability of a farm being given up.
– The same can be said about higher prices farmers receive for food.
– A higher wage level, however, increases the decline in farm numbers.

Generally, economists like to show that people behave in accordance with rationality, and they were able to do this in the case of structural change. Certainly, if farmers could earn more, with more fields and better monetary conditions, they would be more likely to keep the farm. As soon as opportunities outside agriculture became available and attractive, however, the likelihood of abandoning the farm rose.

But what about age, a variable that always proves to be highly significant if included in the studies? The obvious explanation—the older the farmer, the more likely he is to close the farm—is somewhat simplistic. If farms are abandoned, it would be untypical, at least in most countries, for a person to switch from being a farmer to being a driver or a doctor. In fact, a Swiss farm programme providing funds for such re-education had to be closed because only a handful of farmers had subscribed in all the years of its existence. The typical person abandoning his farm is 60–70 years old and enters retirement rather than a new job. And the farm will usually not be abandoned if there is a son (or perhaps a daughter) who is willing to take it over.

This completes our mental journey from inter-generational relations towards economists' concern about structural change in agriculture. In essence, structural change in agriculture is a story about successful or failed farm successions. And these successions usually occur between the different generations on a farm. In order to illustrate this system, it may be useful to enter the world of theoretical modelling. While this requires a degree of abstraction, it may clarify the very relationship between farm succession and structural change.

Let us first focus on determinants of the personal decision to take over a farm. In line with Rosen's model (1986), we assume two kinds of jobs to choose between:

$$u_{ia} = W_{ia} + n_i \tag{2.1}$$

$$u_{ib} = W_{ib} + n_{ib} \tag{2.2}$$

The agricultural job (set equal with taking over a farm) a and the non-agricultural job b; both have two utility components, a monetary welfare measure w that mirrors the amount of money as earned income and a non-monetary utility component n. The non-monetary utility components of farming have been described extensively (Bahner 1995).

It is reasonable to assume that potential farm successors are only a finite number of people M. For this model, M may, for example, be assumed to consist of farmers' children. A broader definition of M would include every school graduate who would prefer to work outdoors.

The decisive factors in M's occupational choice are now the differentials, rather in expected wage W_{ie} than in real wage between agricultural (a) and non-agricultural (b) occupations

$$\Delta W_{ie} = W_{iae} - W_{ibe} \tag{2.3}$$

and in expected non-monetary utility components

$$\Delta n_{ie} = n_{iae} - n_{ibe}, \tag{2.4}$$

so that it is possible to work out the expected difference in utility Δu_{ie} between the two occupational choices

$$\Delta u_{ie} = \Delta W_{ie} + \Delta n_{ie} \tag{2.5}$$

as a result.

Graduates will only choose to enter farming (D = 1) if that is what maximises their expected utility. Otherwise, they will choose the non-agricultural occupation D = 0. Consequently, choices are wholly covered by the rule:

$$\text{Choose } D = 1 \text{ or } D = 0 \text{ as } \Delta u_{ie} \gtreqless 0$$

Ties ($\Delta u_{ie} = 0$) are broken by random device, such as flipping a coin.

Given the size of M choosing between D = 1 and D = 0, relative market supply conditions are completely characterised by calculating the number for whom $\Delta u_{ie} > 0$ and the number for whom $\Delta u_{ie} < 0$. It is convenient to describe differences in preferences among M parametrically for analysis. Define $g(\Delta u_{ie})$ as the density (in the sense of a probability density function) of expectations in the population of M making choices and define $G(u_{ie})$ as the cumulated density. Then, the fraction of M who choose D = 1 must be

$$M_1^s = \int_0^{\infty} g(\Delta u_{ie})du = 1 - G(0) \tag{2.6}$$

The remaining fraction of M chooses not to enter farming. These are persons for whom. $\Delta u_{ie} < 0$, so

$$M_0^s = \int_{-\infty}^{0} g(\Delta u_{ie})du = G(0) \tag{2.7}$$

Figure 2.3 illustrates Eqs. (2.6) and (2.7) for a given distribution of Δu_{ie}. Relative supply to D = 1 farm successors is the area under $g(\Delta u_{ie})$ to the right of 0—this is Eq. (2.6). Relative supply to D = 0 is the area to the left of 0—this is Eq. (2.7). E shows the conditional expectations for the whole group of M as well as for M_0^s and M_1^s.

Finally, the share s of M that engages in farming is defined as

$$s = M_1^s / M. \tag{2.8}$$

Our theoretical considerations in this model lead to the first hypothesis that the expected difference in utility between an agricultural career and a non-agricultural career influences the decision between farm succession and an alternative career.

In order to draw clear conclusions from the patterns of occupational choices to the patterns of structural change, it is convenient to come up with two additional simplifying assumptions.

The first assumption is that the period of being the farmer in charge on a farm is given as t years and does not vary over time. t is assumed to be identical for all farmers. The second assumption is that no exit from the farm household is possible before year t once the decision to take over (D = 1) has been made. Both assumptions do largely

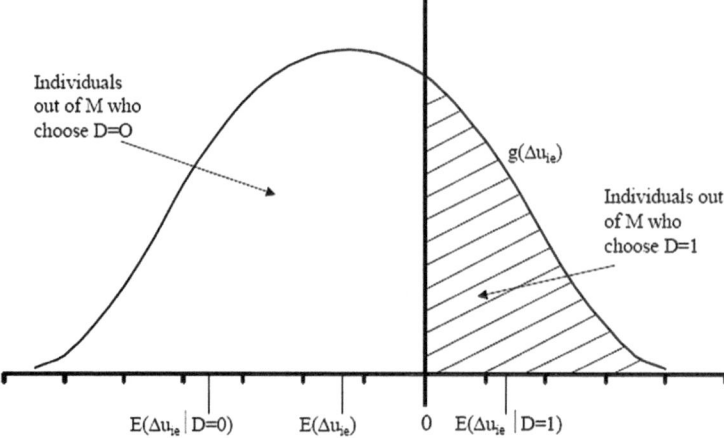

Fig. 2.3 The choice of entering farming

match the empirical results for family farming, particularly full-time family farms, which will be presented in Sect. 2.4. The second assumption can theoretically be explained by the prohibitive level of sunk costs because of investments in education and experience that cannot be regained when leaving the farm.

From here, the number of farms in a given area can be estimated as

$$F = \sum_{j-1}^{t} (s_j * M_j), \tag{2.9}$$

in which $j = 1$ describes the past year, $j = t$ the year after which farmers are going to retire. The rate of structural change in agriculture is on the whole primarily described by the annual rate of variation in farm numbers $\Delta F/F$. This rate can be quantified as

$$\Delta F/F = (s_0 * M_0 - s_t * M_t)/\sum_{j-1}^{t} sj * Mj, \tag{2.10}$$

That leaves two causes of structural change. The first is that $M_0 \neq M_t$, i.e. that the number of persons eligible to take over a farm has changed over the years. Consider farm successors as constituting M. A past decline in the number of farms then decreases M_0 compared to M_t. That makes structural change a self-accelerating process. For a broader definition of M, the demographic decline in the birth rate that was experienced in most of the industrialised world led to $M_0 < M_t$. Structural change in agriculture should therefore be considered in the context of past sociodemographic trends.

The second constituting component for structural change is the size of s. It is therefore worthwhile to specify Eq. (2.10) by inserting Eqs. (2.8) and (2.6).

$$\Delta F/F = \left(\int_0^\infty g(\Delta u_{ie0}) du / \int_0^\infty g(\Delta u_{iet}) du \right) / \sum_{j-1}^t sj * Mj \qquad (2.11)$$

For the current situation, the distribution of the once expected utility of retiring farmers may be assumed as given. t is also assumed as a constant. Figure 2.4 therefore illustrates the rate of structural change as a function of $E(\Delta u_{ie})$ in year $j = 0$. It shows how rational expectations connected with an agricultural career, weighed against rational expectations connected with a non-agricultural career, influence structural change. To give an extreme example: imagine that the expected utility of farming in the current year is so low that nobody enters farming.

Under the assumptions of the model, the maximum rate of farm decline would be restricted to

$$\Delta F/F_{min} = -s_t * M_t / \sum_{j-1}^t sj * Mj \qquad (2.12)$$

Equation 2.12 may be visualised with help of figures. Given that farmers have a period of being in charge on a farm for $t = 30$ years, and given that, in past years, exits from and entries to farming have been constant from year to year, the maximum decline in farm numbers in the current year would be 3.3%.

Point A in Fig. 2.4 depicts a situation in which $s_0 * M_0 = s_t * M_t$, where the number of entries equals the number of exits t years ago, so that the annual rate of structural change is zero. Point B mirrors a situation that is more typical for Western societies. The expected utility of taking over a farm is low, thus not all farms do find a successor. This leads to a decline in the number of farms. Corrado et al. (2017), for example, point to the fact that, between 1990 and 2010, the average farm size in Italy rose from 5.6 to 8 ha, while the number of holdings declined from 2.7 million to 1.6 million. This is a very typical example for industrialised countries today. Point C shows the opposite situation that is typical for some developing countries (Mandal 2000) and for many periods in mediaeval times (Abel 1962). The expected opportunity costs of farming are so low that the number of entrants exceeds the number of exiting farmers, therefore the number of farms increases and the size of the average holding decreases.

It is widely believed that exogenous changes influence occupational choices. The impact of economic changes on structural change can therefore be seen as an indirect connection. Figure 2.4 shows a situation in which agricultural policy conditions change in a favourable way, be it through introduction of direct payments or through an administered increase in food prices. This increases the mean of Δu_{ie}, so that B is shifted towards B'. However, an increase in opportunity costs, for example through an increase in non-agricultural wages or a reduction in unemployment, may again decrease Δu_{ie} and shift the equilibrium back to B. Thus, the speed of negative structural change increases again, as fewer graduates choose a farming career.

It is possible to verify this model empirically. In times when agriculture does better than other sectors, a larger proportion of young people enter an agricultural

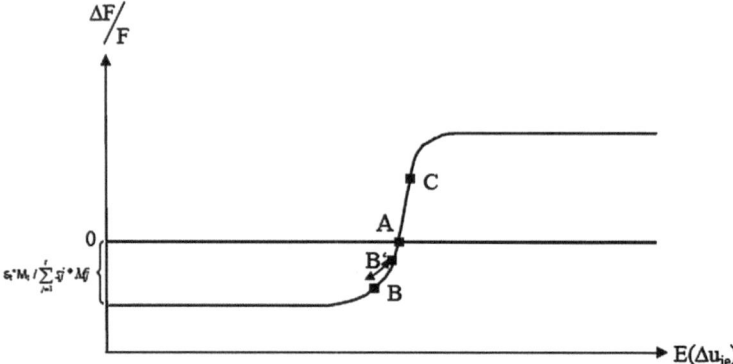

Fig. 2.4 Rational expectations and structural change

education. And the more young people enter such education, the more farms can (and will) be taken over a few years later.

While this is how economic rationality plays into inter-generational relations, nobody would claim that rationality is all there is in inter-generational relations. It is obvious that culture and values play strongly into this issue, both from the older and the younger generation's side. The latter will be explored in more depth in the next section. For the moment, some words about the perspective of the old farming generation about to retire should suffice.

Psychologists have often described the desire, once life is approaching its end, to leave something behind (e.g. Cooper et al. 2009). Yes, everybody's stay on this planet comes to an end. But couldn't I leave something behind that lasts? To just participate a tiny bit in eternity on earth?

For farmers, it is likely that this "something", in the ideal case, is a well-organised, profitable and sound farm. Economically, retired farmers would usually be best off if they sold their land to the neighbours, providing a nice additional income. Empirical results, however, show that preferences of retiring farmers usually lie elsewhere:

– Farm succession often has a lot to do with honour. If it is possible to hand over the farm to the next generation, this is usually an act of joy and pride, even if it happens outside the family (Mann 2007).
– It is even possible to economically trace preparations for a successful succession to the next generation in the family-farm life-cycle. Mann et al. (2013) show that, in the years before retirement, farms that are going to be transferred experience a boom in investment activity. In farms without succession, however, this is the time when disinvestment already begins.

These are some of the few common patterns of farm succession and other inter-generational issues. Others are much more dependent on the respective cultural norms and, even in a single macro-region such as Central Europe, they can strongly diverge. Consider the question of where the retired farming couple would live after succession.

In an identical survey carried out in Switzerland and Northern Germany (Mann and Rossier 2007), 82% of Swiss respondents stated they would continue to live on the farm. 76% of German respondents, however, stated they would move away.

2.4 Concluding Thoughts on Hierarchy

The public administration is a convenient case to become familiar with the structure of hierarchies. Max Weber's "iron cage" is a powerful image to illustrate the unconditionality of hierarchies: if you want to stay in the system, you have to obey, no matter what!

Research in the 100 years since Weber's death, however, has contributed to softening the iron cage by asking naughty questions. Would the organisational units really need the expanding budget they claimed to need? Were they able to control what they claimed to control? Could the organisational structures be adapted to organisational patterns in businesses? And could the costs the administration was causing be justified?

Businesses dealing with each other are already a lecture in "hierarchies for the advanced". Economic models start to get complicated as soon as power asymmetries are to be considered. However, power asymmetries in the chain are the rule rather than the exception in the economy. Or have you ever wondered why your insurance puts you on hold when you call them, though you wouldn't do that to them when they call you, let alone why they never actually call you?

Finally, hierarchies in the family will be all too "familiar" to most of us. However, these hierarchies are extremely dynamic. Apparently, we should be kind to our kids in family businesses, because later on today's small pests will be responsible for the very fate of the business.

By now, readers may have formulated a naughty question of their own: what does all of this have to do with agriculture? More than with, for instance, the educational administration, the health sector or families running a taxi business? Well, many findings of this section can be transposed to other sectors. Agriculture is just a worldwide melting pot with an extremely long tradition that has collected many, and significant, cultural habits. And as the food trade is arguably the most essential backbone of every economy, the farming sector serves as an ideal case in point for the significance of socioeconomics.

References

Abel W (1962) Geschichte der deutschen landwirtschaft: vom frühen Mittelalter bis zum 19 Jahrhundert. Stuttgart: Ulmer

Abbink J (2011) 'Land to the foreigners': economic, legal, and socio-cultural aspects of new land acquisition schemes in Ethiopia. J Contemp Afr Stud 29(4):513–535

Allen J (2007) Integrating life, work and identity: farm women transforming 'self' through personal struggle and conflict. In: Billett S, Fenwick T, Somerville M (eds) Work, subjectivity, and learning. Springer, Heidelberg

Allen DW, Lueck D (2003) The Nature of the Farm. Cambridge: MIT Press

Allers MA (1994) Administrative and compliance costs of taxation and public transfers in The Netherlands. Groningen

Bahner T (1995) Landwirtschaftliche Betriebsgestaltung nach persönlichen Zielen. Agrarwirtschaft 44(10):343–350

Balisacan AM, Sombilla MA, Dikitanan RC (2010) Rice crisis in the philippines: why did it occur and what are the policy implications. In: Dawe D (ed) The rice crisis: markets, policy and food security. Routledge, London

Baxter J (2010) Wie Gold, nur besser. Le Monde Diplomatique, 15 Jan 2010

Bektasoglu B (2012) Gender-role in decision-making. Markgraf, Weikersheim

Berger P, Luckmann T (1966) The social construction of reality. Penguin, Harmodsworth

Bertoni D, Cavicchioli D (2016) Farm succession, occupational choice and farm adaptation at therural-urban interface: the case of Italian horticultural farms. Land Use Policy 57(4):739–748

Cernea M (1978) Macrosocial change, feminization of agriculture and peasant women's threefold economic role. Sociologia Ruralis 18(2–3):107–124

Cooper D, Moodley J, Zweigenthal V, Bekker J-G, Shah I, Myer L (2009) Fertility intentions and reproductive health care needs of people living with HIV in Cape Town, South Africa: implications for integrating reproductive health and HIV care services. AIDS Behav 13(Suppl 1):38. https://doi.org/10.1007/s10461-009-9550-1

Corrado A, de Castro C, Perrotta D (2017) Cheap food, cheap labour, high profits: agriculture and mobility in the Mediterranean. In: Corrado A, de Castro C, Perrotta D (eds) Migration and agriculture. Routledge, London

Cudjoe G, Breisinger C, Diao X (2010) Local impacts of a global crisis: food price transmission, consumer welfare and poverty in Ghana. Food Policy 35(4):294–302

Dawson S (2015): Coke's zero tolerance for land grabs proves difficult to fulfill. http://www.reuters.com/article/us-landgrab-coke-idUSKBN0ML0LE20150325 (17 Feb 2017)

Department for Environment, Food and Rural Affairs (2014) Red tape challenge. DEFRA, London

Dries L, Swinnen J (2004) Foreign direct investment, vertical integration and local suppliers: evidence from the Polish Dairy Sector. World Dev 32(9):1525–1544

FAO (2013) The female face of farming. FAO, Rome

Fährmann B, Grajewski R (2013) How expensive is the implementation of rural development programmes? Eur Rev Agric Econ 40(4):541–572

Fan JPH, Huang J, Morck R, Yeung B (2014) Institutional determinants of vertical integration in China. J Corp Finan http://dx.doi.org/10.1016/j.jcorpfin.2014.05.013 (20 Feb 2017)

Finan A (2007) New markets, old struggles: large and small farmers in the export agriculture of coastal Peru. J Peasant Stud 34(2):288–316

Fletcher AJ, Kubik W (2017) Introduction—context and commonality. In: Fletcher AJ, Kubik W (eds) Women in agriculture worldwide. London: Routledge

Gembreselassie S (2012) Production and marketing of vegetables among smallholders in Ethiopia: the case of Lume district of Ethiopia. In: Africa Farm Management Organisation: The AFMA Congress. Moi University Press, Eldoret

Gorin V, Dubied A (2011) Desirable people: identifying social values through celebrity news. Media Cult Soc 33(4):599–618

Granovetter M (1985) Economic action and social structure: the problem of embeddedness. Am J Sociol 91(3):481–510

Grüske KD (1991) Zur Bürokratieverlagerung im Steuerwesen – Kosten der Steuern innerhalb und ausserhalb der öffentlichen Administration. Das öffentliche Haushaltswesen in Österreich 32(1–2):43–62

Higgins TJ, Fenrich (2011) Legal Pluralism, gender, and land access in Ghana. Fordham Environ Law Rev 7(1):7–29

Hubbard M (1995) The 'new public management' and the reform of public services to agriculture in adjusting economies: the role of contracting. Food Policy 20(6):529–536

Jager R (2004) The fate of family farming. University Press of New England, Hanover

Jensen M, Meckling W (1976) Theory of the firm. Managerial behavior, agency costs, and ownership structure. J Financ Econ 3(4):305–360

Klein B, Crawford RG, Alchian AA (1978) Vertical integration, appropriable rents, and the competitive contracting process. J Law Econ 21(2):297–326

Lowder SK, Skoet J, Raney T (2016) The Number, size, and distribution of farms, smallholder farms, and family farms worldwide. World Dev 87(4):16–29

Mandal MAS (2000) Private sector as an emerging institution for accelerating growth in Bangladesh agriculture. Presentation at IAAE-Conference, 16.8.2000, Berlin

Mann S (2001) Zur Effizienz der deutschen Agrarverwaltung. Agrarwirtschaft 50(5):302–307

Mann S (2002) The concept of administrative elasticity. Int J Pub Adm 25(8):1007–1019

Mann S (2003) Theorie und Empirie agrarstrukturellen Wandels? Agrarwirtschaft 52(3):140–148

Mann S (2007) Understanding farm succession by the objective hermeneutics method. Sociologia Ruralis 47(4):369–383

Mann S, Rossier R (2007) Nationale Unterschiede und Gemeinsamkeiten bei der Hofübergabe im deutschsprachigen Raum. In: Kuhlmann F, Schmitz P (eds) Good Governance in der Agrar- und Ernährungswirtschaft. Landwirtschaftsverlag, Münster

Mann S, Mittenzwei K, Hasselmann F (2013) The importance of succession on business growth: a case study of family farms in Switzerland and Norway. Yearb Socioeconomics Agric 2013:103–131

Mengel C, von Cramon-Taubadel S (2014) Distance and border effects on price transmission: a meta-analysis. http://hdl.handle.net/10419/97323 (13 Feb 2017)

Muenstermann I (2010) Too bad to stay or too good to leave? Two generations of women with a farming background—what is their attitude regarding the sustainability of the Australian Family Farm? In: Luck GW, Black R, Race D (eds) Demographic change in Australia's Rural Landscapes. Springer, Heidelberg

Mughal MM, Akram M (2011) Does market size affect FDI? The case of Pakistan. Int J Contemp Res Bus 2(9):237–247

Muslim A (2011) Elasticity of corn price transmission and its implication to farmers. Econ J Emerg Markets 3(1):77–85

Nave-Herz R (2013) Ehe- und Familiensoziologie. Beltz, Weinheim

Niamir-Fuller M (1994) Women livestock managers in the third world. IFAD, Rome

OECD (2015) Innovation, agricultural productivity and sustainability in The Netherlands. OECD, Paris

Oxfam (2017) The truth about land grabs. https://www.oxfamamerica.org/take-action/campaign/food-farming-and-hunger/land-grabs/ (15 Feb 2017)

Pollitt C (1993) Managerialism and the public services: cuts or cultural change in the 1990s?. Blackwell, London

Puppin-Lerch S (2007) Contemporary African women artists. ProQuest, Ann Arbor

Raab U (1995) Öffentliche Transaktionskosten und Effizienz des staatlichen Einnahmesystems. Berlin

Rørstad PR, Vatn A, Kvakkestad V (2007) Why do transaction costs of agricultural policies vary? Agric Econ 36(1):1–11

Rosen S (1986) The theory of equalizing differences. In: Ashenfelter O, Layard R (eds) Handbook of labor economics, vol I. Amsterdam, Elsevier

Rossier R (2004) Familienkonzepte und betriebliche Entwicklungsoptionen. Agroscope, Tänikon

Sandford C, Godwin M, Hardwick P (1989) Administrative and compliance costs of taxation. Fiscal Publication, Bath

Schulze B, Spiller A (2006) Is more vertical integration the future of food supply chains? In: Bijman J et al (eds) International agri-food chains and networks. Wageningen Academic Publishers, Wageningen

Tran M (2013) Coca-Cola vows to axe suppliers guilty of land grabbing. Guardian 8(11):2013

Tremblay F (2008) Why parenting is invalid. https://francoistremblay.wordpress.com/page/263/? pages-list (13 Jan 2017)

Tuong V (2009) Indonesia's agrarian movement: anti-capitalism at a crossroads. In: Coaouette D, Turner S (eds) Agrarian angst and rural resistance in contemporary Southeast Asia. Routledge, London

Vavra P, Goodwin B (2005) Analysis of price transmission along the food chain, OECD Food, Agriculture and Fisheries Papers, No. 3, OECD Publishing, Paris. http://dx.doi.org/10.1787/ 752335872456 (14 Feb 2017)

Weber M (1905) The protestant ethic and the spirit of capitalism. Penguin, London

Weldegebriel HT (2004) Imperfect price transmission: is market power really to blame? J Agric Econ 55(1):101–114

Williams JC (1989) Deconstructing gender. Mich Law Rev 87(4):797–845

Chapter 3
Agricultural Markets

Since the times of Adam Smith, ten thousands of economists have devoted most of their efforts to understanding markets. The resulting narrative they have helped develop impressively explains a large number of real-world phenomena.

The most basic version of this narrative goes as follows: Two selfish individuals find that they can increase their utility by trading with each other. They engage in the trade on a fully informed base and end up happier than they were before.

Economists also acknowledged that life can be somewhat more complex than that and have developed their models accordingly. For example, they modelled cases where full information was not available to one or both of the market partners. All these extensions of the basic model have led to a large body of thought about exchanges. Millions of pages have been filled to adapt the two selfish individuals to the linkages and insecurities in the real world.

In this chapter, we will examine three market-related fields of decision-making in the realms of agriculture and rural areas. We start by coming back to the issue of occupational choice, this time mainly from a market perspective. A second section will then focus on the dynamics of settling or depopulating rural areas from a market perspective. Finally, we will look at choices on the food market.

3.1 Occupational Choices

The chapter on hierarchies (Chap. 2) already reflected on the issue of occupational choice. However, few other decisions have such a strong and lasting effect on the course of our lives as the choice which professional path to follow. Moreover, few other decisions are as important for the scope of agriculture as we know it today.

We therefore will broaden the reflection on occupational choice by two aspects. First, we will analyse how a young person decides whether to enter the parents' farm. However, farm work is not necessarily a lifetime decision, nor does it mean to take over a farm. Therefore, we will look at motivations for employed and temporary farm

© The Author(s) 2018
S. Mann, *Socioeconomics of Agriculture*, SpringerBriefs in Economics,
https://doi.org/10.1007/978-3-319-74141-3_3

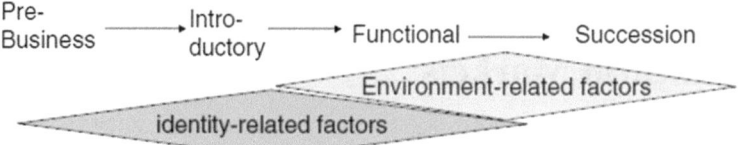

Fig. 3.1 The process of occupational choice (Mann 2007a: 437)

work in a second step. The insights we gain will result—once again—in a theoretical model.

3.1.1 Identity and Environment

When modelling the process of occupational choice in the previous chapter, we took a rather abstract 'economic' approach. We assumed a preference for or against becoming a farmer among the potential successors, instead of concluding such preference from an analysis. In fact, economists tend to assume preferences as given and leave it to psychologists to focus on the genesis of preferences.

One of the most thorough empirical analyses of the complex interplay between parents' expectations and their children's ambitions, of the roles of gender and personal character, was done 20 years ago in New Zealand by Keating and Little (2007). They found that '*parents are less aware of the messages they are sending to their children than are the children themselves*' (p. 167). Parents define their choice, or recommendation, based on gender, health and abilities.

It is important to note that the process of succession has several phases and that each phase has different driving factors. Psychologists emphasize that adolescence is the phase of identity generation. Youth is a time in which longer-lasting preferences and abilities have to be defined. More often than not, the passage from adolescence to early adulthood then involves the individual adjusting to the environment. The 'storm and stress' years of identity craving are followed by self-establishment and adaptation.

All of these factors result in a theoretical model of farm succession as depicted in Fig. 3.1. Identity-related factors in the farm succession process are a liking for working outside, for working with animals or for working self-employed. Environment-related factors are the farm's potential to generate income, the prospects of agricultural subsidies or tax laws. Along the different phases of the succession process, it is likely that environment-related factors gradually become more important, whereas the role of identity-related factors decreases over time.

Some years ago, a quantitative study from Switzerland (Mann 2007a) tested and partly confirmed this model. Seven hundred and thirty-one children of farm managers aged between 14 and 35 years returned a questionnaire on their intention to take over the parental farm. The responses showed that agreement with sentences like 'I like working outside' decreased in their predictive power with increasing age. However, although the importance of environment-related factors could be shown, it was not possible to verify empirically that their impact increased with increasing age.

One fringe result of the study referred to the gender aspect (Mann 2007a). The farm managers' daughters did not only respond in fewer numbers than the sons did, they also showed much less inclination to take over their parents' farm. The interpretation of this finding is left to the reader: Are women intrinsically less interested in farming than men are, or did the parents' behaviour hold daughters back from becoming engaged in agricultural affairs?

3.1.2 Part-Time Farming

Part-time farming is an arbitrary term. If the farmer's spouse earns a few Euros per year by helping in the church, will the farm be considered a part-time farm? Or does this term require a farm manager who earns at least 50% of the income off-farm?

Off-farm income is much easier to define than part-time farming. Furthermore, one of the certain long-term tendencies in worldwide agriculture is the rising share of off-farm income. In the US, nowadays 90% of the total income of farm households comes from sources outside agriculture (Briggemann 2011).

This trend means, of course, that one of the presumptions in the previous subsection was misleading if not false. Becoming a farmer is not really a binary decision. For example, if you are unsure whether to become a teacher or farmer, you have the option of becoming both. Many social scientists in the past thought that part-time farming would merely be an intermediate stage during a family's phasing out of agriculture. At least for the short and medium terms, this view could not be confirmed. Empirical studies (e.g. Kimhi 2000) have shown that off-farm income may even stabilize the persistence of farms.

Nonetheless, disadvantages of part-time farming exist. At least in countries with small farms, such as Switzerland, part-time farms and other small-sized farms suffer from missing economies of scale and therefore low degrees of productivity (Mann 2007b). The economic disadvantages of part-time farming were also demonstrated for Norway: Mittenzwei and Mann (2017) found that incomes of households with 100% farm income or 100% non-farm income exceed the income of households with shared incomes.

Despite these economic aspects, there may be good arguments for pursuing a part-time strategy. Tzouramani et al. (2014), for example, showed that Greek farmers enter off-farm activities to buffer their risk. Work satisfaction may also play a role: Carrying out rather specialized and monotonous tasks is less satisfying than being engaged in a broad range of activities, in agriculture and beyond (Besser and Mann 2015). Thus, a trade-off between monetary and non-monetary utility may exist, a thought that we will address later in this chapter.

Such trade-off certainly applies to the 5% of Swiss farms that generate a negative agricultural income, which the farmers usually compensate by a positive (and often generous) off-farm income (Mann 2015a, b). These farmers choose to manage their farm for non-financial reasons. The lines between work and leisure are certainly blurring in such cases.

3.1.3   Employed on a Farm

We have focused so far on the decision to become the manager on a family (full-time or part-time) farm. While this is a decision that 500 million active farm managers (this is the number of family farms as estimated by Lowder et al. 2016) have made at some stage in their lives, an even larger number of persons have made the decision to be hired on a farm, a group that this book so far has neglected. Two important differences between the two groups are immediately obvious:

– Farm work as an employee is a much shorter-term decision than becoming a farm manager, even in countries in which becoming a farm manager is often not a lifetime issue. Farm workers are often hired for months, weeks or even days.
– Although the literature is ambiguous on the merits of becoming a farm manager, social scientists do not draw a beautiful picture of the conditions and lives of farm workers. Being a farm worker is often considered as a precarious occupation.

This latter point can be illustrated from a historical and from a systematic perspective. To start with the former, the history of slavery in the US is a very visible case for showing how the least-privileged people—African Americans—mostly ended up as farm workers. Even in 1900, decades after the abolishment of slavery, more than half of all farm workers were still African Americans (Perry et al. 2014). Now as then in most societies, immigrants are still the people living under precarious conditions, and they provide a lot of the workforce of major farms in wealthy countries, be it Mexicans in the US, Romanians in Germany or Moroccans in Spain. Repeatedly, riots break out, as Corrado et al. (2017) showed in an international record.

Generally, one says that although farm managers have to work hard for an income that is on average rather low, they enjoy a great degree of freedom and still have the chance to make a lot of money on the volatile agricultural markets. Farm workers do not share these advantages. By receiving low payments (Zhao 1999) and carrying out monotonous work delegated by their managers (Pierson 1978), they do not participate in the bright sides of farming.

This situation implies a great need for unions to represent the interests of this vulnerable segment of the workforce. Paradoxically, the degree of organization among farm workers is extremely low. Saverio Caruso (2016: 277) explains: '*Among the key reasons for labourers' weak position in relation to their employers are the predominant seasonal and informal nature of labour relations, the lack of direct links between workers and employers, and, consequently, the difficulty in organizing collective bargaining mechanisms.*'

3.1.4 Alpine Summers

In the previous subsection, we described an environment in which labour is hard, workdays are long and payment is low. Could it possibly get worse? What if the labour market existed for only three months? What if you knew beforehand that you have to look for another job to cover the remaining nine months of the year? What if—to make things worse—housing and consumption choices were particularly basic?

This is what happens on Alpine summer farms. Nevertheless, between 30,000 and 50,000 people work there each year, and most of them would have other opportunities; in fact, some of them even leave well-paid jobs with banks or consultancies to enter Alpine summer farming.

For readers less familiar with the habits in Europe's largest mountain area, we will outline the basics of this peculiar labour market: Around one million hectares of mountainous grassland are covered by snow for a good time of the year. However, in the time between June and early September, grass grows that can be utilized for grazing. Therefore, farmers in the region often choose to send their cattle, their sheep or (occasionally) their goats to summer farms where other people watch their animals, herd them and milk them. This custom saves the farmers not only feedstock but also labour that can be used on their land.

Part of the Alpine summer farms are run by single farmers, others by cooperatives or local authorities. In any case, they usually hire staff for which they provide some basic housing in the middle of nowhere. Accessibility varies, and in many places, food is brought from the lowlands just once a week.

A Swiss–Italian researcher (Calabrese et al. 2012, 2014) visited 50 Alpine summer farms in Switzerland and interviewed 120 workers to understand the underlying motivations. A cluster analysis showed that the workers could be categorized in four groups:

- 'Tourist workers' were the group matching the understanding of labour markets in which one primarily worked to make a living. This group was mostly composed of male foreigners, who viewed low salaries in Switzerland as being not too bad by international standards. While these workers often reported conflicts with their employers, they appreciated contacts with tourists on the summer farm.
- The situation was completely different for the 'eremites', who worked on Alpine farms to get away from everything and enjoy the solitude. They usually had a non-agricultural background and mostly were faithful workers, likely to attend training beforehand and to return to their summer farm in the next year.
- For the 'nature lovers', the motivation for entering the experience was not so much the longing for solitude but the appreciation of the landscape, nature and animals. This aspect turned out to be not the best motivator for this kind of job: Nature lovers were likely to come into conflict with their colleagues and to leave Alpine farming behind as a one-time experience.
- By far the oldest age group were the so-called 'traditionalists', for whom coming to the summer farm was a fixed part of their yearly routine. They considered Alpine farming as a duty and complained about their inexperienced colleagues. Usually,

they were male and Swiss, and often they managed the summer farm jointly with their family.

For employers, it is inconvenient if a new team arrives every summer requiring them repeatedly to explain all work steps. As everybody wishes the workers to return in the next year, Calabrese et al. (2014) carried out a regression analysis to identify both financial and non-monetary factors with a predictive power for the employee's intention to return in the next year.

It sometimes can be illuminative if certain variables do not have any explanatory power. This was the case for all financial variables involved. Employees were asked whether they found their wage level satisfactory, they were asked about their wage level, and, dependent on education and experience, their wage was compared with a wage that could have been expected. None of these three variables did have any predictive power for the intention to return to the farm in the next year. Apparently, money is not the key motivator for Alpine farming.

Two other factors were much more helpful in explaining decision-making. One of them was infrastructure. It mattered, for example, whether the living quarter had a heating system. Being cold at night made a worker much less likely to return to the place thereafter. The other factor was conflict, be it with the employer or in the team. Rude remarks and arguments apparently were no good ingredients for a rich experience in the Alps.

3.1.5 The Activity Choice Model

Economic science has occasionally been criticized for a biased view on labour, which often is reduced to its money-generating function although it is only one side of the coin. Pagano (1985: 173) summarized this shortcoming with the following simple claim:

> Conservative priests used to prescribe the status quo by saying that life itself was a means to a superior end existing somewhere in the sky; economists would assume a similar role by maintaining that working life is simply a means to a superior end, existing somewhere on earth, called consumption goods and leisure. But our working life affects our welfare as much as our non-working life and the availability of consumption goods.

The model by Rosen (1986) as outlined in Chap. 2 at least acknowledged that two important aspects influence the process of occupational choice: money and non-monetary utility. The empirical evidence presented in the last two subsections underlined this notion, adding two important insights:

- Occupational choice may be, but rarely is, a once-in-a-lifetime decision. Although we still face a large degree of path-dependency in our careers, more and more people have to decide what to do next more than once in their lives. Therefore, the category of occupational choice should be supplemented with the category of activity choice. Activity choice would allow me some time of 'work & travel' on a

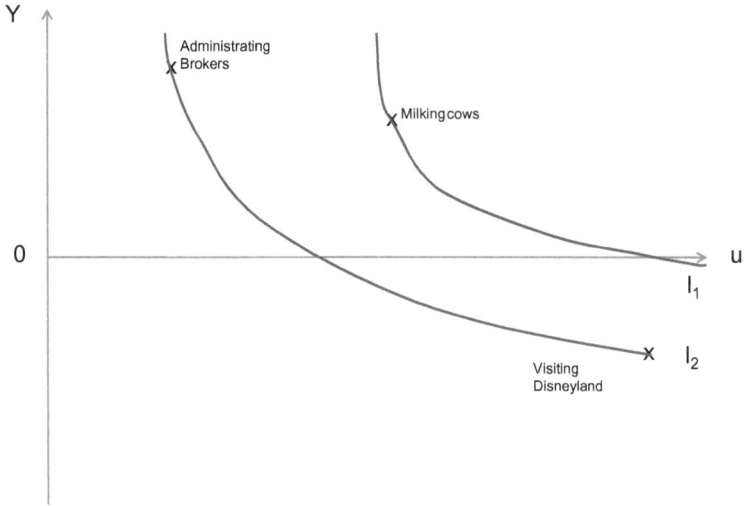

Fig. 3.2 The activity choice model

Canadian farm and a later job as a university teacher in Switzerland, which I then may interrupt by a summer of Alpine farming.

– The trade-offs between earnings and non-monetary utility may vary considerably. The case of Alpine summer farming demonstrates instances when the monetary component exists but shrinks in comparison with non-monetary utility components.

All these considerations lead to the 'activity choice model' as depicted in Fig. 3.2. It acknowledges that every activity has a monetary component as represented along the y-axis. An activity (such as chatting with my neighbour) may involve no money exchange so that it would be situated on the u-axis where $y = 0$. Other activities with $y < 0$ (going shopping, eating out) are typically allotted to consumption, whereas activities with $y > 0$ are usually considered as work.

The u-dimension describes the non-monetary utility of an activity. Boring or even disgusting activities will be situated on the left side, and the curves shift to the right as occupations become more interesting and attractive.

The person considered in Fig. 3.2 can choose between three alternatives at a certain point of time: Administrating brokers is apparently an unattractive activity but generates a high income. Milking cows seems to generate more enjoyment but less financial return. The most non-monetary utility can be generated by visiting Disneyland, but this activity is linked to spending money.

For a rational choice between these three options, an additional piece of information is needed: the relationship with which the person in the model trades non-monetary utility against money. As usual in the world of microeconomics, this relationship is illustrated by a set of two indifference curves. As people favour a lot

of money and fun over little money and fun, indifference curves on the upper right
are preferred to indifference curves on the lower left so that

$$u\,(I_1) > u\,(I_2)$$

Translated back into practical terms, this equation implies that the individual
would prefer milking cows for a small income to both working as a broker with a
higher wage and visiting Disneyland with a higher fun factor.

In general, the activity choice model is a convenient tool to demonstrate how
activities with low monetary returns can be a perfectly rational choice. It has been used
as an argument against a minimum wage in which activities in the range minimally
above the u-axis (i.e. with a low, but positive hourly return) are outlawed (Mann
2014a).

In the realm of agriculture, the activity choice model also has been used to illustrate
the dual way in which external crises affect agricultural structures. Sadly, Australian
agriculture could serve as a convenient example, because the country's farmers have
faced many crises: In the early 21st century, many years of severe drought made
large areas of land completely unproductive for years. Nevertheless, some regions
were also affected by temporary flooding. Furthermore, Australia has hardly any
protective measures such as tariffs on its borders. Thus, economically, farmers face
the whole range of price volatility present on the world market, meaning that prices
for their products went far down on several occasions.

Shaped by culture and by necessity, Australian farmers are among the most flexible
in the world. Different to, for example, their European colleagues, they are usually
ready to sell their farm and buy another or change to an entirely different occupation.
On that base, the framework in the previous chapter would suggest that it was likely
that crises such as drought, flooding or price drops would reduce the profitability of
farming and therefore make it more likely that farmers opt out.

Coming back to the activity choice model, the model structure suggests that our
view of the Australian circumstances may have neglected a second dimension, the
dimension of work satisfaction. When crops stop growing, cows die of thirst, fields
are flooded before harvest or when wool prices collapse, how will such circumstances
affect the joy of managing a farm? If I am not able to successfully raise animals and
grow corn, can I stay happy as a farmer? Will crises make it less likely that I carry
on farming even if I ignore the economic aspect?

Structural equation models are a good tool to check for such complex interrelations
because variables can be both cause and effect in a system. A structural equation
model as depicted in Fig. 3.3 verified the two main drivers causing structural change:
happiness and profitability. Although crises affect happiness and profitability more
or less equally, happiness seems to surpass profitability in its effect on planned farm
exits.

Summarizing, occupational choices always have an economic and a cultural,
identity-related component. Both are worth studying. We presume that this interplay of economic and psychological factors for decision-making also exists in other
areas where decisions have to be made. Let us examine them.

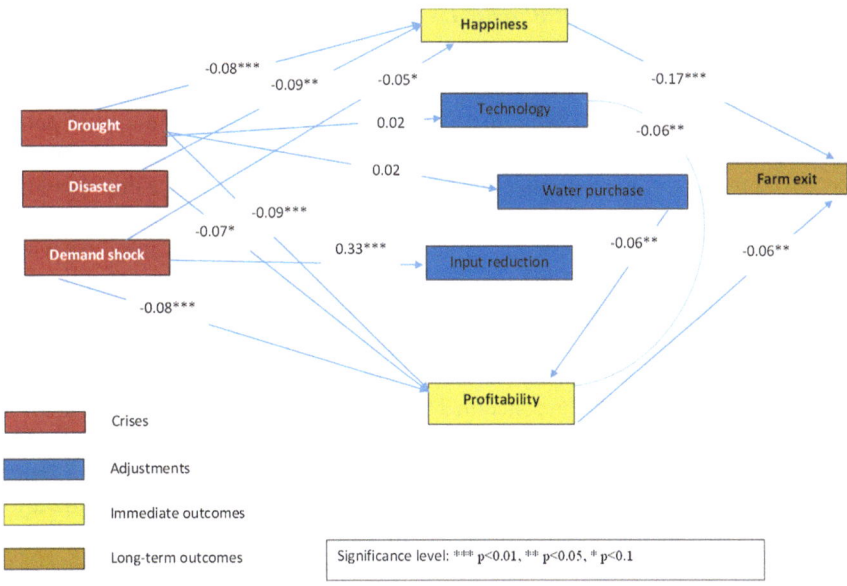

Fig. 3.3 Structural change by profitability and happiness in Australian agriculture (Mann et al. 2017)

3.2   Residential Choices

Once you made the decision to take over the farm of your parents, you are usually spared at least one other decision: where to settle down. However, as a reader of this book, you are more likely to face this decision at one or even several stages in your life: Are you looking for a job in a big city or in the countryside? Are you going to move abroad or stay in your country? Will you commute between your home and your work place? The following pages should clarify some of the linkages between this micro-level and the macro-level of regional development.

However, the following subsection will focus on the primary sector and neglect urban issues. Urbanization has been a worldwide trend in the last centuries, as cities, for many people, have apparently become an increasingly attractive option for spending one's lifetime. This trend is partially responsible for the depopulation of peripheral areas. Choosing a clear focus on rural areas with their traditional agricultural character in the context of this volume leads us to the issue of rural development.

For the farming sector, this is important because a sustainable development of rural areas is often cited as one of the reasons to promote or at least keep farming enterprises in peripheral areas which would depopulate and degenerate otherwise. Therefore, the link between the primary sector and rural development should receive special attention over the next pages.

3.2.1 The Tiebout Model and Its Limits

For economic science, the 1950s arguably were a period of significant intellectual development. Robert Samuelson, in 1954, developed a formal definition of public goods. Although it had been clear that certain items would have to be delivered by the government, this mental model clarified under which conditions this should be the case.

The two and a half pages in his 'Pure Theory of Public Expenditure' on which Samuelson laid out these conditions proved to be an important impulse for many scholars to build on these ideas. One of them was Charles Tiebout, then a PhD student at the University of Michigan. He placed an article in the Journal of Political Economy called 'A pure theory of local expenditures' (Tiebout 1956). Whereas Samuelson had mainly been concerned with public goods on the national level, Tiebout thought about goods as being delivered by local authorities. He drew the picture of a well-functioning competition between the local communities in a region or country. There would be towns that would charge high taxes and provide a lot of infrastructure such as good schools, public baths and parks, and there would be towns with lower taxes but without the public amenities. In this model, no need exists for national authorities to interfere because people effectively vote on their feet. If I do not like poor infrastructure, I move to the town with higher taxes, and vice versa.

If this model is taken seriously, there is no need for regional policies. If people find their local development problematic, they are free to move to another region. Peripheral rural counties may die out, but this is how markets work, and likely, people can find regions in other parts of the world that could fit their preferences much better than their present location.

However, research in the last 50 years unearthed much criticism of the Tiebout model, both on theoretical grounds and by empirical experience. These repeated criticisms can be summarized in the following three points:

- Mobility comes at a cost. Although it is possible to move from one community to another, the cost of paying for the moving truck with the furniture is just a minor part of the involved costs. When moving over larger distances, people will have to look for a new job, and the effort to build up a new social network is considerable, particularly for elder people.
- Before moving can begin, costs arise to obtain necessary information. Most countries have thousands of municipalities. It will be impossible to check for all of them (let alone for foreign ones) whether their infrastructure and their local policies suit my needs. Information costs may become prohibitively high.
- Villages need a certain minimum size to finance some necessary infrastructure such as schools, a size that Rieder (1991) estimated as 500 inhabitants for Swiss conditions. At the other end of the spectrum, some economists find that certain megacities have already exceeded an economically reasonable size (Moomaw and Alwosabi 2004). Tiebout, for whom each municipality had an optimal size, had ignored both scenarios.

These critical voices should not keep us from appreciating Tiebout's contribution to the field: He opened an extremely important debate on how regions should deal with processes of immigration and depopulation. It is likely (as often) that extreme solutions will not lead to an optimal outcome. Neither should artificial infrastructures be established and maintained in regions in which nobody wants to live, nor should we fully trust the market to fix problems and prevent regions from dying out, leaving only the old and vulnerable behind whereas the young have moved to more promising regions.

3.2.2 Policy and Politics of Rural Development

Over the last 50 years, policy makers have kept an increasing distance to Tiebout's approach. Both within and outside agricultural policy, the emphasis on rural development in the policy portfolio has certainly increased, particularly in European countries. Mainly two approaches have emerged over recent decades and have been taken by most programs.

For one, it is obvious that not all municipalities, no matter how small, can offer all amenities one could imagine. Nevertheless, it is essential that people can easily access infrastructures such as schools and hospitals, maybe even theatres and swimming pools. To solve this contradiction, policy makers as early as in the 1960s began to focus on a few central locations in problematic regions. In countries with declining population, such as Germany, this 'approach of central locations' helped to prevent outmigration; in countries with high population growth, such as India, it helped transform small towns into vivid growth centres.

The second tool is support for the establishment of enterprises in peripheral regions. Some governments emphasize the relocation of large, existing companies, whereas others see more potential in convincing employed or unemployed people to found their own business.

Such policies, pursued in peripheral areas, may have their origin in two different political frameworks. They may be part of

– General economic policy: Officials in charge of improving entrepreneurship and economic growth may feel responsible for creating instruments designed for poor and peripheral regions.
– Agricultural policy: Officials in this realm may find it too narrow to support just farms for rural development and may want to broaden their support to other sectors.

In some countries, responsibilities have been allocated to one of the two. In other countries, some programs are run by economic and others by agricultural authorities. The latter scenario applies to many regions in the EU. A thorough analysis of such parallel activities was carried out for Switzerland (Mann 2014b), where very similar rural development programs have been in place in parallel for many years, with the economic authorities distributing 30 million Swiss Francs each year and the agricultural authorities 7 million. Actors on the local and national levels were interviewed to understand their motivations. On the local level, the analysis identified lack of information exchange as the main problem. Often, actors involved in key positions were not aware that similar programs under a different umbrella would support similar projects. On the national level, actors were much better informed and organized regular exchanges. However, they were careful not to question the equilibrium that policy makers had obtained and were inclined to keep the status quo.

3.2.3 Turning Points in Village Developments

To this point, we have only addressed the question of political visions of rural development. Especially peripheral regions clearly need the attention of policy makers, which they actually receive in most countries. There is a public interest, in theory and practice, to keep these areas afloat.

What we still need to discuss is the rural development in the real world. We cannot take for granted that rural development policies will indeed fulfil their promise. To do so, they would have to have an impact on residential choices. The programs would have to influence, directly or more probably indirectly, the decisions of families and individuals where to move or how many children to get. These two variables —migration and birth/death ratios—are the two only instruments having a direct impact on population development.

Usually, population development tends to be extremely stable over long time spans. Geographers have remarked that most regions that have lost inhabitants over the last decades are very likely to do so in the future and vice versa: Areas with increasing population in the past are likely to attract immigrants in the future.

As a first step towards understanding motivations and mechanisms of residential choice, Mann and Gennaio (2011) looked at 10 Swiss villages in which this pattern was disrupted. In five of them, population figures had gone up over decades until, between 1960 and 1990, this development reversed and people emigrated. The other five villages had a clear emigration tradition over decades until, also between 1960 and 1990, people started to move inwards.

A similar analysis could be carried out for cities. For example, Dresden (Germany) and Tallinn (Estonia) both had turning points around the turn of the century, when a negative migration trend reversed into a positive one. However, a city is of course much more complex than a village with a few dozen inhabitants such as Vals in the Swiss Grison Canton. For Vals, it is probably easier to produce a consensus about the reason behind the turning point than for Tallinn.

Well, Vals was perhaps not the best example because it produced somewhat of an outlier reason for its increase in population after a long history of population decline. Residents chose to settle in Vals after a number of companies had invested in this remote mountain location, including a refurbishment of the mineral water factory by Coca Cola Inc. This circumstance was not typical for a village experiencing a positive turning point in its population development. The other four villages in Switzerland—Vorderthal, Doppleschwand, Sant Antonio and Schwändi—portrayed a different scenario. Each used to be a farming village with a steadily decreasing number of farms. At a certain point, when urban employees became willing to commute over longer distances, they considered living in these villages and commuting to work. In addition, urban people began migrating into the villages, building houses they would not be able to afford where they worked, and driving to work in the nearby towns each day.

The only similarity between these villages and those with a turning point towards population decline was the existence of one outlier among the five case studies. The outlier in the latter was the small village of Kirchenthurnen, where development was planned on a small piece of land on which construction of new housing should take place. However, a series of bankruptcies prevented the construction projects. Spatial policies in Switzerland did not allow a second piece of land to be converted to building land as long as the first parcel was still available. Here, we can speak of bad luck that entailed a declining number of people in the constant number of houses in the village.

However, the other four cases—Linthal, Andermatt, Airolo and Trun—were probably more interesting from a scientific viewpoint. They used to be either flourishing industrial villages or villages with a large military base. In Linthal and Trun, the downturn of the Swiss textile industry in the second half of the 20th century led to a number of shutdowns and considerable job losses. Who would produce Swiss clothing if trousers and shirts could be imported from China for a fraction of the price? At the same time, the cold war was ending, and military bases in Airolo and Andermatt were reduced considerably. The reduced number of jobs led to a reduced number of residents.

These two sets of stories show some asymmetry. Depopulation could be traced back to structural issues in the local economy. Population boosts, however, could usually not be explained by the local economic performance but rather by the provision of parcels for urban commuters. Economics, after all, might become less important over time to explain rural dynamics, at least in the sense of local production activities.

3.2.4 Core Factors for Rural Areas

Case studies are always convenient for an illustration, but larger sample sizes provide much more certainty in terms of general patterns. For this purpose, the population development of villages was analysed econometrically for two regions, namely Switzerland (Mann 2004a; Mann and Erdin 2005) and northeast Germany (Mann 2004b). Thus, some facts are available about the driving forces of residential choices and birth rates.

The studies showed that the number of agricultural jobs had a negative impact on population development in northeast Germany with its few commercial farms. For Switzerland with its many small family farms, the opposite was found. Furthermore, the number of farms (irrespective of their size) in a region was strongly correlated with the region's birth rate.

These puzzling differences can be explained by the different systems and their different rationales. Northeast German farms used to be greatly overstaffed during socialist times. Since then, an ongoing process of saving on jobs has made regions in which farming played a major role prone to unemployment and subsequent outmigration. In Switzerland, the population level is very stable. However, whereas an average Swiss woman gets 1.4 children over her life span, an average Swiss farmwoman gives birth to 2.7 children. This difference can best be explained by transaction costs. On Swiss family farms, parents spend most of the time engaged in activities where they can supervise their children relatively easily, if compared with 'classical' office careers. Easy child supervision is much less the case on commercial farms of several thousands of hectares in size. Therefore, small-scaled family farming systems have a positive impact on the birth rate.

Another clear difference between the two rural systems is the effect of unemployment. For northeast Germany, we see an expected negative effect of unemployment on population development. The higher the unemployment in rural counties, the higher is the outmigration. In Switzerland, the effect has the opposite direction. In many Swiss regions, unemployment is frictional, i.e. short-term. In these regions, persons entering phases of unemployment do not decide to move away but patiently wait until the next job opportunity in the region opens up. As a result, there may be a positive correlation between unemployment and population growth.

However, when combining economic and non-economic variables into explanatory factors, the predictive value of non-economic factors is stronger than that of economic variables. A high share of people above 65, combined with a low share of young people, is a powerful predictor of low birth rates. Good accessibility by public transport surprisingly correlates with a negative migration balance. Apparently, trains and buses are mainly used to leave the region rather than to immigrate. The factors summarizing traditional agricultural structures, a high presence of industrial enterprises and a strong presence of the service sector do not have any positive impact on population development. The factors regarding the farming and industrial sectors even have a significant negative impact on the migration balance.

On the municipal level, these relationships imply that we more or less deal with a post-work society. For a rural region, attracting new businesses is partly uncorrelated and partly even negatively correlated with attracting new residents. From a hedonistic viewpoint, it is understandable that neither industrial enterprises nor pig or poultry stables create the most attractive environment for building and inhabiting a new residential home. However, the outlook of villages increasingly having to choose between becoming a production hub or a residential place raises difficult questions for rural development policies. Their focus on attracting businesses may become outdated, and new approaches will have to be identified.

3.3 Production and Consumption Choices

Of all subjects covered in this book, the subject of production and consumption choices is certainly the one with the most literature available. In the many textbooks and articles about farm management, one can find a lot of reflection on which crops to grow and which animals to raise. Regarding consumption, the subject concerned with consumer choices can be distinguished into demand analysis and agricultural marketing. Whereas demand analysis focuses on aggregated data with a particular emphasis on prices, agricultural marketing focuses on the unique characteristics of single products and the best strategies for increasing sales.

This short section will of course not be able to compete which this abundant and inspiring body of literature. We will have to be very selective, focusing within this rather economic topic on subjects with an increased relevance of social factors. In fact, we will focus on the choice to decline possibilities. This section starts by highlighting the organic movement, a movement that decided not to make use of artificial fertilizers and synthetic pesticides. It then proceeds to genetically modified organisms (GMO), or rather to segments of the population choosing not to consume GMO-products. Finally, we will look at consumers opting for non-consumption for other ethical and cultural reasons.

3.3.1 Organic Production and Consumption

For centuries, foodstuff on the shelf was distinguishable by the eye, for example apples and pears. One of the arguably biggest innovation brought about by producers and marketers was the use of labels to distinguish identically looking products, such as Golden Delicious apples, depending on whether they were organic or conventional apples. Although brand segregation had existed before, the concept of raw product segregation was born.

Actually, the concept of segregation has two aspects, a technical and a cultural one. It is important to distinguish between these two aspects.

Technical segregation is an engineering-based task, making sure that different product groups, even if visually not different, remain reliably segregated until they reach consumers' refrigerators. Organizational measures have to be established that prevent lower-priced conventional food to be labelled as organic food. In some dairies, for example, first organic milk is bottled, then conventional milk, and then the plant is cleaned. That way, spilling of conventional milk into organic milk is prevented, while spilling of organic milk into conventional milk does not cause any problems. Some organic producer associations do not allow farms to produce partly organically and partly conventionally because they are concerned about technical segregation.

Cultural segregation emphasizes the political and philosophical contents of the organic movement. Many scholars and members of the organic movement have labelled organic production as anti- or post-productivist and as an ecological movement against multinational food firms. A typical practical example of cultural segregation is local organic retailers. The decline of their importance in many countries, linked with an increasing coverage of the organic segment by the large international retailers indicates that the importance of cultural segregation is declining. This observation has often been termed the 'conventionalization' of organic farming.

As discussed in the preceding sections, most farms are taken over from the parent generation, and their organic or conventional production status will have followed a 'path-dependency'. Most farmers, at some stage in their career, will consider a system change, even if only for seconds. In fact, an increasing number of farms have changed from organic to conventional production (Sahm et al. 2013), not only in the opposite direction.

Many scholars in the past used the concept of innovation theory to explain the conversion from conventional to organic (e.g. Padel 2001). As already mentioned, the main characteristic of organic farmers is the rejection of certain production factors. Both mineral fertilizer and pesticides have only been available for a —historically—relatively short period. There are many good reasons not to make use of these innovations. However, reasons to view this rejection as an act of particular innovation are rather sparse. Moreover, as the rate of organic farmers starts to exceed 10% of total farmers in several countries, the concept of innovation theory becomes increasingly less applicable.

Lamine and Bellon (2009) provided a valuable review of the literature on the conversion towards organic farming, distinguishing normative from comprehensive studies and studies concerned with the short time of conversion itself from studies taking into account the longer process of preparation and adaptation. They see a particular lack of attention to this latter point, as few social scientists have taken the time to observe long-time processes of transformation.

This point is very important because the farms switching from one system to another only represent a certain group of farmers, termed 'optimizers' by Mann and Gairing (2012). Usually, these converters use their production system as a tool to generate a decent income. Farmers remaining in their system—be it organic or conventional—are very loyal to their current system. A large number of conventional farmers would never consider converting to organic production, and many organic farmers cannot imagine allowing conventional production factors on their farms.

This distinction between 'loyals' and optimizers can also be made for consumers. Nevertheless, the number of consumers who never eat any organic food is small in most developed countries, and even smaller is the number of consumers who never eat anything that is conventionally grown. In fact, for consumers, 'optimizing' is the usual behaviour. In this process, some kind of selectivity can be observed: We find more consumers choosing organic lemons and carrots but conventional beer and chocolate than consumers choosing the opposite.

A Swiss study (Götze et al. 2016) showed that consumers have fixed patterns regarding what products they preferably buy in organic quality. Most importantly, the stronger the degree of processing the lower is the share of organic food. This choice is not rational because both the health-related and the environmental effects of organic compared with conventional food do not depend on the degree of processing.

It shows that our intentions are different when we buy raw produce as compared with our purchases of ready-made pizza.

Another finding was that the market share of organic products is higher for imported foodstuff than for domestically produced farm outputs. This choice is more rational than the one just described. Apparently, the organic label serves as a guarantee for certain production standards, as consumers may know that legal standards differ considerably from (exporting) country to country.

Organic farming is not gender neutral. Among the relatively few female farmers in western countries, the share of organic farmers is always higher than among male farmers, as for example Jacobson et al. (2003) showed for Florida and Bjorkhaug (2006) for Norway. The same can be said about the consumers. All analyses of consumption data show that women are much more likely to purchase organic qualities than are men, from Australia (Lockie et al. 2002) to Ireland (Davies et al. 1995).

The level of education also influences producers and consumers similarly in their preference for organic agriculture. Most studies find that '*regular consumers tend to be educated*' (Padel and Foster 2006: 606), and studies on the adoption of organic farming find that better educated farmers produce organically more often than less educated farmers (e.g. Mzoughi 2011). Again, education works symmetrically on the production and consumption side.

The idea that producers and consumers of organic food share not only common values but also social characteristics is not new (Storstad and Bjorkhaug 2003). However, in recent years sufficient evidence has accumulated: Producers and consumers of organic food constitute an influential social group. The social aspect may very well be a significant driver in the success of organic farming. Hundreds of production systems have been promoted, including integrated production or animal-friendly production. Nonetheless, only organic farming, relying on a few simple rules, has been able to establish in almost all countries in the world and to obtain a high degree of public attention.

3.3.2 Genetically Modified Organisms

After scientists had made technical advances in agriculture, a segment of consumers decided they would not want to take advantage of them, and a segment of producers decided the same. This plot sounds familiar from the preceding subsection on organic farming, and it seems to duplicate itself in the story of genetically modified (GM) crops. Thus, we deem it worthwhile to examine both the parallels and the differences between the two stories.

The time dimension is certainly among the strongest differences. Artificial fertilizers were synthesized in the 19th century, and synthetic pesticides started their career in the early 20th century. Genetic modification, however, became possible from around 1980 onwards.

Another difference is the extent of proven harm the technologies have been causing. The US Fish and Wildlife Service estimates that 72 million birds are killed by pesticides in the US each year. Fatalities for humans are certainly lower, but world-wide deaths and chronic diseases may reach one million per year. GMO certainly do not reach this extent of acute damage. However, the opponents of GMO raise three major points against working with modified seeds:

– About 80% of GM crops have herbicide resistance as their innovative character-istic. This means that the main impact of GMO is the increased usage of total herbicides. The 'Non-GMO Project' argues that the use of total herbicides like Roundup has increased by the factor 15 since GMO came on the market.
– For pesticides, some detrimental long-term effects only became visible decades after usage. It still can be argued that negative long-term effects of GMO (be it on human health or the environment) may not be visible yet, although this argument loses weight over time.
– GMO-technology requires major investments in infrastructure and knowledge. This is an advantage for multinational companies like Monsanto and Syngenta and prevents their use for minor breeding companies. Therefore, the use of GMO-technology has implications for the seed industry's structure.

These (and other) reservations against GMO have led a number of countries such as Russia, Peru, Venezuela or Austria to ban the use of GMO in agriculture and even imports of GMO-produced food and feed. Other countries such as Germany and France also do not allow cultivation but tolerate imports. This is another significant distinction to the organic market, where no government has banned the cultivation of conventional crops or the import of conventional food products. The reason for this difference cannot be the harm caused by GMO—or conventional production but more likely lies in the benefits. Few governments would like to impair the productivity that artificial fertilizers and synthetic pesticides have brought to their agriculture, which would be the consequence of banning conventional agriculture. Compared with this scenario, the costs of banning herbicide-resistant GM crop varieties are relatively low.

There is also a political reason that would speak for banning GMO, at least for European administrations. According to a 'Eurobarometer' in 2010, 59% of Euro-peans believe that GM food is not safe for their health. This high degree of scepticism leads us to the question of labelling. If food containing GMO is sold, should con-sumers be made aware of it?

Two agricultural economists from Greece developed a model after their emigration to the US showing that labelling brings advantages. Their model, as shown in Fig. 3.4, sorts consumers in relation to their attitude towards artificial technologies. On the very left, you find consumers indifferent to how their food was produced. On the very right, consumers are opposed to any kind of artificial inputs and have a high preference for organic food.

In the reference scenario, there is no labelling of GM food. The net utility for eating non-labelled (nl) food is defined by the gross utility (U) and the food's price (p_{nl}), as is the utility of eating organic (o) food, which depends on U and p_o. Predictably, the x_{nl} segment on the left derives higher utility from eating non-labelled food, whereas

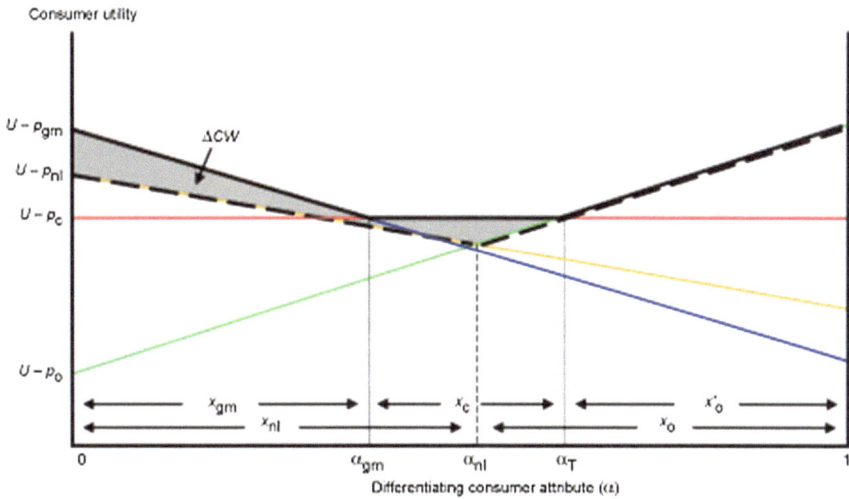

Fig. 3.4 Market and welfare effects of mandatory labelling of genetically modified products (Giannakas and Yiannaka 2006)

the x_o segment right of α_{nl} will choose organic food because utility is higher than when consuming the non-labelled food.

After mandatory labelling, consumers can distinguish between GM food with a net utility of $U - p_{gm}$ and conventional (c) food with a net utility of $U - p_c$. Figure 3.4 shows how this scenario affects market shares: Organic food actually loses market shares as only the x'_o segment still relies on organic food. Between the extremes, there is a segment x_c that purchases conventional food, whereas the x_{gm} segment buying GM food is able to cut costs by buying labelled GM food. This effect of cost saving and the benefit of being sure that the food does not contain GMO causes the two grey areas, which represent the positive consumer welfare effect (ΔCW) of mandatory GMO-labelling in the model. This model neglects, of course, transaction costs, which are always on the negative side of mandatory labelling.

Whereas this model treats labelling of GM food as an either–or issue, the reality is—as usual—more complex. In Germany, for example, labelling is mandatory if food contains parts of GM crops (almost non-existent on the German food market). However, labelling is not mandatory for animal products that were produced with GM feed, and these products are very common.

In milk, for example, traces of GM feed used during production may not be detectable. Nevertheless, there is a demand among consumers for milk produced without GM feed. As no mandatory labelling exists for this kind of milk, some producers use negative labelling to advertise milk that was produced without GM feed. This market segment for 'GMO-free milk' can be divided into three parts:

- Because organic production outlaws GM feed, the market for organic milk is also a market for GMO-free milk. Producers of organic milk receive a premium of around 8 Cents per litre.
- Some dairies use the label 'ohne Gentechnik' (without GMO) and contract farmers who do not use GM feed. To these farmers, they pay a premium of 0.3 to 1 Cents per litre.
- Other labels like Haymilk or fair milk link GMO-free feed with some other environmental or social attributes, usually paying premiums between the two figures above.

This overview indicates that the attribute 'GMO-free' has not at all attained a similarly convincing power as the organic label. However, market niches have been created.

It is worthwhile to compare the outlined German approach with the Swiss one. The legal situation in this adjacent country is the same: It is legally possible to import GM feed, although the cultivation of GM crops is not allowed. However, market partners in Switzerland stopped importing GM soybeans in 2007, after a blackmailing campaign by Greenpeace. Since that time, Swiss feed importers have shouldered some effort and additional costs to import GMO-free soybeans, mostly from Brazil. Of course, from a welfare-oriented point of view, this strategy causes losses for consumers who face higher-than-necessary milk prices if they do not prefer GMO-free milk.

The global market for soybeans and corn (by far the two economically most important GM species) shows that technical segregation can occur on three levels:

- We find countries (such as Switzerland) that segregate on a national level by excluding GM feed, but a larger number of countries have a market for GM feed plainly because there is no demand for GMO-free qualities. Portugal would be one example.
- Segregation can also take place within a company. Some logistic enterprises, for example, have trucks they use for GM feed and others they use for GMO-free qualities.
- The most frequent level of segregation, however, is the company. Particularly multinational companies such as Cargill and Bunge offer only GM qualities, whereas slightly smaller companies such as ACTI and COFCO often offer exclusively GMO-free qualities.

The theoretical model depicted in Fig. 3.5 may explain why the company as segregation level is chosen so often. The more centrally segregation is carried out the higher are frustration costs. This scenario applies, for example, to Swiss consumers who would like to buy milk containing GMO if they can save a few Cents by doing so. The technical costs, however, are high if segregation is carried out within companies. Likely, minimum total costs occur at a level (L_{min}) between these extreme solutions.

Fig. 3.5 The costs of
technical segregation (Mann
2015a, b)

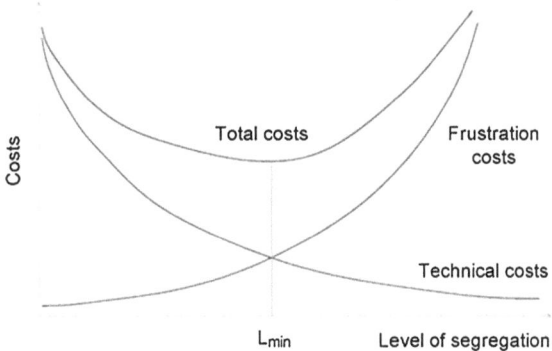

3.3.3 More on Non-consumption

So far, we have shown that concerns for the environment or for one's health can be reasons to forgo certain consumption options such as conventional or GM food. However, a growing number of people renounce meat (and often fish) consumption neither for health nor for environmental reasons. They consider it as wrong to kill animals for the purpose of eating them and thus choose to eat a vegetarian diet.

The best country to study vegetarianism is certainly India, where 30 to 40% of the people do not eat meat. The 360 million vegetarians in India probably outnumber vegetarians in the rest of the world. The reason for this unique situation is religion. It would be an undue simplification to say that Hinduism prescribes vegetarian diets, but the least one can say is that Hindus pursue a particularly complex relation between caste, class, gender, kinship and meat consumption. Fischer (2016) nicely illustrated how this social acceptance of vegetarian diets relates to food markets. It had led, for example, to labels for vegetarian products and unusual product portfolios of otherwise standardized fast-food chains such as McDonalds.

If consumers do not want animals to die for them, vegetarian diets that contain eggs and milk may not do the job. More often than not, the fate of male chicks in egg production and of male calves in milk production is not apt to high ethical standards. Therefore, proponents of vegan lifestyles (where you avoid any product of animal origin) certainly have a strong argument to make. However, although vegan products are among the fastest growing segments on the food market, people who eat strictly vegan diets still represent less than 1% of the population in most countries.

An obstacle to measuring the share of vegetarian, vegan and even organic diets is methodology. Sociologists are aware that respondents to surveys do not necessarily tell the truth but like to tell things they expect society to appreciate, a motivation called 'social expectancy'. An illustrative example is organic food, for which the share of organic products bought according to surveys was around 20% in Germany at times when the real market share was around 2% (Simons et al. 2001). For vegetarian and vegan diets, 'market shares' are much more difficult to measure. It is likely that not all of the 13% of Swiss people stating vegetarian diets in a survey are indeed vegetarian.

An analysis of the purchasing behaviour of households over one month showed the share of vegetarian households in Switzerland to be below 5%. Nevertheless, many Swiss consumers have concerns about the ethical justification of eating meat, although they are not able or willing to switch to a vegetarian or even vegan diet (Berndsen and van der Pligt 2004).

Hinduism, by the way, is not the only religion that supports vegetarianism. In Christianity, vegetarian diets were prescribed for centuries in the weeks before Easter. In most Christian countries, this habit has been fading. Today, forgoing chocolate or alcohol, if anything at all, has become as popular as forgoing meat. In Romania, however, a significant part of the population still follows the 'Lent' practices of the Orthodox Church. This practice forbids the consumption of any animal product with only very few exceptions and of alcohol. In recent years, while Romanians have gained some material wealth when compared to either the dire years of socialism or the difficult period of political-economic transformation, the practice of fasting gains some attractiveness again—the share of households not consuming alcohol and animal products during Lent is rising.

References

Berndsen M, van der Pligt J (2004) Ambivalence towards meat. Appetite 42(1):71–78
Besser T, Mann S (2015) Which farm characteristics influence work satisfaction? An analysis in two agricultural systems. Agric Syst 141:107–112
Bjorkhaug H (2006) Is there a female principle in organic farming? In: G. Holt, M. Reed (eds) Sociological perspectives on organic agriculture. CAB International, Wallingford
Briggemann BC (2011) The importance of off-farm income to servicing farm debt. Federal Reserve Bank of Kansas City. Econ Rev (First Quarter):63–82
Calabrese C, Mann S, Dumondel M (2012) Patterns of occupational choice in the Swiss alpine labor market. Yearb Socioecon Agric 2012:31–54
Calabrese C, Mann S, Dumondel M (2014) Alpine farming in Switzerland: discerning a lifestyle-driven labor supply. Rev Social Econ 72(2):137–156
Corrado A, de Castro C, Perrotta D (2017) Cheap food, cheap labour, high profits: agriculture and mobility in the Mediterranean. In: Corrado A., de Castro C, Perrotta D (eds) Migration and agriculture. Routledge, London
Davies A, Titterington AJ, Cochrane C (1995) Who buys organic food? A profile of the purchasers of organic food in Northern Ireland. British Food J 97(10):17–23
Fischer J (2016) Markets, religion, regulation: Kosher, halal and Hindu vegetarianism in global perspective. Geoforum 69(1):67–70
Giannakas K, Yiannaka A (2006) Agricultural biotechnology and organic agriculture: national organic standards and labeling of GM products. AgBioForum 9(2):3
Götze F, Mann S, Ferjani A, Kohler A, Heckelei T (2016) Explaining market shares of organic food: evidence from Swiss household data. Br Food J 118(4):931–945
Jacobson SK, Sieving KE, Jones GA, Van Doorn A (2003) Assessment of farmer attitudes and behavioral intentions toward bird conservation on organic and conventional Florida farms. Conserv Biol 17(2):595–606
Keating NC, Little HM (2007) Choosing the successor in New Zealand family farms. Family Bus Rev 10(2):157–171

Kimhi A (2000) Is part-time farming really a step in the way out of agricultural? Am J Agr Econ 82(1):38–48

Lamine C, Bellon S (2009) Conversion to organic farming: a multidimensional research object at the crossroads of agricultural and social sciences. A review. Agron Sustain Dev 29(1):97–112

Lockie S, Lyons K, Lawrence G, Mummery K (2002) Eating 'Green': motivations behind organic food consumption in Australia. Sociologia Ruralis 42(1):23–40

Lowder SK, Skoet J, Raney T (2016) The number, size, and distribution of farms, smallholder farms, and family farms worldwide. World Dev 87(4):16–29

Mann S (2004a) Landwirtschaft und ländlicher Raum. Agrarforschung 11(2):44–49

Mann S (2004b) Die Entsiedelung ländlicher Räume und das Agrarsystem. Berliner Debatte INITIAL 15(2):46–55

Mann S (2007a) Tracing the process of becoming a farm successor on Swiss family farms. Agric Hum Values 24(4):435–443

Mann S (2007b) Zur Produktivität der Nebenerwerbslandwirtschaft in der Schweiz. Agrarforschung 14(8):344–349

Mann S (2014a) Individual utility effects of minimum wages in a new activity-choice framework. Forum Social Econ. https://doi.org/10.1080/07360932.2014.946944

Mann S (2014b) Understanding parallel structures in public administration. Int J Org Theory Behav 17(3):275–292

Mann S (2015a) Deconstructing agricultural labour: a reflection on negative farm income. J Socio-Econ Agric 8:70–75

Mann S (2015b) Is "GMO Free" an additional "Organic"? On the economics of Chain segregation. AgBioForum 18(1):4

Mann S, Erdin D (2005) Die Landwirtschaft und andere Einflussgrössen auf die Bevölkerungsentwicklung im ländlichen Raum. Agrarwirtschaft 54(5):258–264

Mann S, Freyens B, Dinh H (2017) Crises and structural change in australian agriculture. Rev Social Econ (accepted)

Mann S, Gairing M (2012) 'Loyals' and 'Optimizers'—shedding light on the decision for or against organic agriculture among Swiss farmers. J Agric Environ Ethics 25(3):365–376

Mann S, Gennaio M-P (2011) Wendepunkte in der Dorfentwicklung. ART-Schriftenreihe, Bd, Tänikon, p 16

Mittenzwei K, Mann S (2017) The rationale of part-time farming: Empirical evidence from Norway. Int J Social Econ 44(1):53–59

Moomaw RL, Awusabi MA (2004) An empirical analysis of competing explanations of urban primacy evidence from Asia and the Americas. Ann Reg Sci 38(1):149–171

Mzoughi N (2011) Farmers adoption of integrated crop protection and organic farming: do moral and social concerns matter? Ecol Econ 78(8):1536–1545

Padel S (2001) Conversion to organic farming: A typical example of the diffusion of an innovation? Sociologia Ruralis 41(1):40–61

Padel S, Foster C (2006) Exploring the gap between attitudes and behaviour: Understanding why consumers buy or do not buy organic food. Br Food J 107(8):606–625

Pagano U (1985) Work and welfare in economic theory. Basil Blackwell, Oxford

Perry N, Reybold LE, Waters N (2014) 'Everybody was looking for a good government job' occupational choice during segregation in Arlington, Virginia. J Urban Hist 40(4):1–23

Pierson R (1978) What about the farm workers. J Agric Econ 29(3):233–242

Rieder P (1991) Die Erhaltung der bäuerlichen Kulturlandschaft der Alpen. http://e-collection. library.ethz.ch/eserv/eth:49257/eth-49257-01.pdf (June 14, 2017)

Rosen S (1986) The theory of equalizing differences. In: Ashenfelter O, Layard R (eds) Handbook of labor economics, vol I. Amsterdam, Elsevier

Sahm H, Sanders J, Nieberg H, Behrens G (2013) Reversion from organic to conventional agriculture: a review. Renew Agric Food Syst 28(3):263–275

Saverio Caruso, F. (2016): Unionism of migrant farm workers. In A. Corrado, C. de Castro, D. Perrotta: Migration and Agriculture: Mobility and change in the Mediterranean area. London: Routledge

Simons J, Vierboom C, Härlen I (2001) Einfluss des Images von Bio-Produkten auf den Absatz der Erzeugnisse. Agrarwirtschaft 50(5):286–293

Storstad O, Bjorkhaug H (2003) Foundations of production and consumption of organic food in Norway: common attitudes among farmers and consumers? Agric Hum Values 20(2):151–163

Tiebout C (1956) A pure theory of local expenditures. J Political Econ 64(5):416–424

Tzouramani I, Alexopoulos G, Kostianis G, Kazakopoulos L (2014) Exploring risk management strategies for organic farmers: a Greek case study. Renewable Agric Food Syst 29(2):167–175

Zhao Y (1999) Labor migration and earnings differences: the case of rural China. Econ Dev Cult Change 47(4):767–782

Chapter 4
Agricultural Cooperation

Cooperative settings, if judged from a utilitarian perspective, can be justified in two distinct ways. The first is that individual utilities can be enhanced if forces are joined together in a cooperative setting, while the second argument concerns utility interdependencies. The fact that my well-being is dependent on my neighbour's well-being not only has added complexity within the mathematical landscapes of utility theory, but has also been used as a theoretical concept to explain practices such as donations (Yandle 1974).

As discussed above, cooperative settings are likely to cause moments of extreme happiness, although such cooperative settings might also result in extreme failure. In this chapter, the incidents of trial and error encountered along the agri-food chain will be investigated and commented on.

4.1 Producing Jointly

There is little doubt regarding the long historical tradition of joint land management. Indeed, the German revolutionary Engels (1882 (1962)) offered an overview of the long history of jointly organising agricultural production over the centuries, exaggerating, perhaps a little, the size of the communities during earlier periods of history in which members would care for each other. His conclusion was that contemporary private farming units, which had been liberated from aristocratic estate owners, would be far too small to be competitive, particularly if America chose to enter global food markets.

Less normatively, a question arises concerning why land management should have become less and less collective over the centuries. Engels explained this by means of political pressure, with an aristocratic elite using wars and laws to expropriate farmers, thereby exploiting them with regards to their accumulated holdings. Economists prefer a different explanation. They cite transaction costs as the main driving factor. If all decisions have to be taken jointly, the decision-making process

© The Author(s) 2018
S. Mann, *Socioeconomics of Agriculture*, SpringerBriefs in Economics,
https://doi.org/10.1007/978-3-319-74141-3_4

requires a significant amount of time and effort, and more efficient management can often be achieved by putting land under the full control of a single skilled producer. The progress achieved by installing such small management units led, for example, to Alpine villages from around 300 AD that managed to supply themselves with sufficient heating and feeding material to sustain the inhabitants over the winter.

What was known during the 20th century as Marxism and what was largely built from a concept that Friedrich Engels helped to develop, led to the top-down organisation of collective land management in many countries. This resulted in cooperation within a hierarchy. In order to know more about cooperation in its pure form, it is more interesting to observe cooperative land management that has emerged through a bottom-up process. Hence, three examples of this process will be presented in this chapter.

While all three examples face their own specific challenges, they do not reflect the numerous attempts at joint agricultural production that have completely failed. An infamous history of such attempts is still waiting to be written and would probably begin with the British entrepreneur Robert Owen (1771–1858), who provided land that he split into "quadratic paradises" on which everybody was welcome to farm jointly. His experiment certainly suffered from the fact that most of the people who were attracted to the idea could not cope with society as it was—and it had to be abandoned after three years, during which time debts had been accumulated and members had become frustrated.

4.1.1 Alpine Grassland

When it was mentioned above that land in Europe originally used to be managed collectively before, over the centuries, it moved into private hands, this was correct for the majority of the agricultural area. Up to the present day, many forests are still owned by a cooperative. For the realm of agriculture, the summer farming areas in the Alps are the exception that proves the rule. Systematic readers of this book will remember the earlier brief encounter with the people who spend three months of the summer travelling into areas in which neither men nor cattle live during the rest of the year, using (once again) the Swiss situation as a case in point. Around half of these people work for private enterprises, while the others are employed by collective enterprises. Some of these collective enterprises are legally part of the municipality, while others are organised as cooperatives.

After previously determining the profile and motivations of their employees, it is perhaps worthwhile briefly explaining the economic rationales of these organisations in only being physically active between June and September. While the costs involved are mainly labour costs, the revenues come from two sources; they charge farmers to look after their animals and they receive transfer payments from the government because of the positive effects Alpine farming has on both biodiversity and the landscape.

In the Valais Canton, Switzerland's most remote area, one of these collective organisations has become relatively well-known among social scientists. Törbel has not achieved this prominence by being more innovative than the adjacent collective organisations, but rather because the anthropologist Netting (1974) happened to choose Törbel as a location for his research and, more importantly, the political scientist Ostrom (1990) used Netting's findings as a case study during her seminal systematic exploration of the realm of cooperation.

The origins of the Törbel cooperative date back to 1293, when one landowner sold four pastures to the local community. The owner retained the right to let his animals graze on the land, so that the financial arrangement was similar to today's mortgages. Unfortunately, we do not know whether and how the land was commonly managed during these early times.

The next information regarding the cooperative dates from the 14th century, when 40% of Törbel's land was commonly managed by nine farmers, while the remainder of the land was in the hands of aristocratic families. At that time, however, the role of the land rents being paid to aristocratic landlords was already in decline.

The formal foundation of the cooperative dates from February 1st 1483, when 22 local farmers signed a joint agreement to better regulate the use of their grassland and forests. One important element of this agreement was exclusion. If external persons were to purchase land in their village, it would not imply the right to use land belonging to the cooperative. This rule has remained fairly typical to date. Most Alpine land cooperatives are not accessible for external persons. Membership remains a privilege for those families who have been members for many generations.

The fact that there were no terms in the agreement concerning conflict resolution indicated that there was a broad acceptance of the terms of the collaboration. The next information source we have is an inventory of the cooperative, which was recorded in 1507. A more comprehensive collection of legal regulations was then introduced in 1517, which have remained valid until now. In particular, a second provision prohibiting the overuse of the grassland was installed. Each member of the cooperative was only allowed to send as many cows to the common land during the summer as he was able to feed through the winter; otherwise, high penalties would apply. Additional prescriptions regarding immigration and emigration, hunting, veterinary control, dispute settlement, village administration and the construction of joint housing were added to the regulations in 1531.

Törbel and many other cases of collective Alpine grassland management that have persisted over the centuries show that this type of land lends itself to the possibility of collective management. The main conclusion Netting (1974) derived from his research concerned the importance of natural conditions. It was no coincidence, he argued, that arable land, intensive meadows and settlement areas were rarely collectively managed, while forests and extensive pastures were more often collectively owned and managed. In fact, Alpine pastures offer several amenities when managed collectively:

– If everybody had to look after their own animals, particularly in the case of herding, labour costs would be much higher.

Table 4.1 Decisive factors for cooperative land use

	Collective land use	Individual land use
Production value	Low	High
Potential for intensification	Low	High
Use frequency	Low	High
Use dependency	Low	High
Yields	Low	High
Area	Large	Small

- There would be a need for a lot of fences and rules of access for remote pastures in the case of private properties. Control costs would also be much higher.
- Avalanches can be better prevented by forests if forest management is jointly planned.
- It is likely that externalities are taken into account more strongly if a larger number of local stakeholders are involved in decision-making process.
- If stones obtained from the land can be used without limitations, then it reduces transaction costs.

All these considerations can be generalised into an appraisal of the merits of collective land use, as depicted in Table 4.1. Based on this appraisal, large, extensively used areas appear to offer more benefits than pitfalls if they are managed collectively.

4.1.2   African Pastoralism

Studying the Swiss case highlighted the significant importance of natural factors. In Switzerland, the political framework has remained extremely stable over the centuries, meaning that the effects of changes in the institutional framework could not be studied appropriately in Switzerland. That does not mean, however, that political forces are unimportant. Pastoralism in Africa, which was described as "the least expensive method of livestock raising" by Konczacki (2014: 168), is a rather convenient example of the importance of this set of factors.

   Mobility is a convenient case in point. The collective use of grassland only works if herds can move freely in order to avoid overgrazing on single spots of lands. However, if the political situation does not allow mobility, it becomes vital to adapt the regime in the best possible fashion. The sustainable management of collectively used land can only be achieved if natural and political conditions are aligned in an appropriate way.

   A significant amount of African grasslands are not the subject of clear property titles and they are jointly managed by local groups. Nevertheless, these groups often allocate well-defined rights of use to their members, frequently taking into account flooding or dry periods.

Some scholars criticise the European perspective for misperceiving African landscapes as natural, while, in fact, they are strongly shaped by cultural influences (Haller et al. 2013). Such misunderstandings hamper a targeted discourse regarding appropriate options for development.

The Kafue Flats in Zambia offer a very well-documented example of the challenges such landscapes face today, challenges that range from a distortion of the natural environment (climate change leading to increased erosion and decreased quality of grasses) to the transformation from a collective to a private institutional setting initiated by colonial and postcolonial powers (e.g. Haller et al. 2013).

Today's tribes have inhabited the Kafue Flats since around 1800. They number around 27,000 persons, resulting in a population density of 18 persons per square kilometre. As the lowland is flooded during certain times of the year, villages were installed on the highland. The tribes originally had a chief who was able to define rules of access to the grassland that were broadly accepted. Spiritual aspects used to play an important role in this process. The spirits of the ancestors were asked to support the contemporary institutional framework, while ritual activities had to be carried out when moving the cattle. Every tribe consisted of several camps, with every camp housing several families who benefited from the land. A camp's coordinator was usually reimbursed with one calf per family per year. His job profile comprised

- settling emerging disputes,
- avoiding conflicts through good planning and
- defending the camp against predators.

In many cases, the camps had agreements with other camps so that mutual grazing would be tolerated.

The Kafue Flats represent a good example of the unintended consequences of measures instituted as a result of good intentions. One of these good intentions was the electrification of the region, for which a large dam was built in order to generate water power. However, the dam worsened the natural conditions of the flats. Bushes and invasive crops had to compete with the grass, which meant that less cattle in a less healthy state could be fed on the land.

Many economists, including those working for the World Bank and the International Monetary Fund, consider clear property rights to be a precondition for economic development. During the 1980s in particular, this led to clear recommendations being made to many governments, including Zambia's, to define property rights. The government complied with such recommendations in 1995 by allocating the right to issue property certificates to the president as well as tribal chiefs. This was intended to generate incentives for investments in the land. In the Kafue Flats, traditionally a stronghold of the opposition, one of the opposition politicians managed to obtain property rights over large segments of the land so that he could charge the local population for tenure. At the same time, a local chief began an irrigation project with the aim of improving the food security of the land. Additionally, as another change, the state introduced inheritance rights, which also supported property rights concerning the cattle.

It is, of course, more difficult to wander around with your cattle if you have to ensure that you do not accidently enter land that you are forbidden to access. The introduction of property rights therefore decreased the pastoralists' mobility considerably. In addition, the new actors started using land that had been traditionally restricted to the native population, partly illegally. Both factors resulted in the overgrazing of the land, thereby causing a decrease in the quality and productivity of the land. As a result of such natural and institutional changes, Merten and Haller (2008) could show a sharply decreased level of food intake among the local population as well as the impaired growth of children.

As Merten and Haller (2008) noted, "the paradox of a state that is simultaneously absent and present" is causing grave distortions in society. The state is present where traditional institutions are dismantled and replaced by new ones, but it is absent where the new framework needs to be implemented in a reliable and sustainable way. After all, pastoralism in the Kafue Flats is an example of a bottom-up cooperative scheme that is being replaced by more market-based settings—with doubtful outcomes.

4.1.3 The Kibbutzim

Although "Israel" may be the initial association when people hear about the Kibbutzim, this way of institutionalising cooperative production is actually older than Israel itself. Jewish settlers, having few chances to organise communities except for joint farming, founded the first Kibbutz in 1909 in Palestine. For Jews in Palestine, that is, before Israel was founded in 1948, the Kibbutzim were more or less the only institutionalisation of Jewish communal life.

This situation changed, of course, when Jews got their chance to run their own state. The young state of Israel attracted a lot of immigrants. In particular, hundreds of thousands of immigrants from Eastern Europe, often with little formal education, ended up in the Kibbutzim as low-cost workers. They soon dominated agricultural production. The 1970s can be considered the heyday of the Kibbutzim as the institutionalisation of joint agricultural production. They combined a rural work environment with a middle-class living style.

Russel et al. (2011) elaborate the challenges that the Kibbutzim subsequently faced, which can be summarised in three main points:

- While the Kibbutzim had been constructed around the notion of joint property and the cooperative management of assets, an increasing number of labour contracts with external persons provided a challenge to the original concept, since there were clearly differences in wealth between insiders and outsiders.
- The cooperative momentum that made individuals forget about their personal standard of living was part of the pioneering spirit of the young state of Israel. It is natural that this pioneering spirit faded over time, and with it, the willingness to live an ascetic life.

- Internationally, the downfall of real-world socialism did not have an encouraging effect on the concept of cooperative and solidary production.

The Kibbutzim survived, however, by adapting to external changes. They started, for example, to acknowledge private property, which was particularly important for many members when it came to their place of residence. They also started to install a differentiated wage system. Further, parts of their cooperative organisations were privatised, for example, educational or health service units. The most consequential step that was sometimes taken was the transformation into a Moshav, which still works the land jointly, but bids farewell to common property outside the realm of production. This may have seemed a step back in terms of cooperative settings, but it did justice to most members' preferences.

4.1.4 Water Management

In some countries, such as the UK or Switzerland, farmers barely think of water as a scarce resource. In most parts of the world, however, water is among the scarcest of resources, particularly in the agricultural sector. Further, if compared to other important factors such as land, seeds or tractors, water usually has far less clear property rights associated with it because it is much more, well, fluid.

One of the least contested scientific findings regarding water management was described a long time ago by White (1957: 160): "If there is any conclusion that springs from a comparative study of river systems, it is that no two are the same." Considering that many farmers use ground water rather than river water as well as the fact that this institutional arrangement of ground water utilisation also shows significant diversity, it becomes clear that a library rather than a paragraph in a socioeconomic textbook would be needed to do justice to the diversity of arrangements concerning water management in different parts of the world. Further, only some of these arrangements are cooperative in nature. In many cases, companies market water to farmers as a simple commodity. In other cases, the state has enacted clear laws regarding how much water farmers are allowed to use, meaning that a hierarchical setting applies. But still, many institutional arrangements are built on cooperation, either partially or fully. This applies at both the local and the international level.

On the local level, a Kurdish colleague once told me that the main motivation for attending Friday prayers at his village's mosque for most local farmers was the opportunity to coordinate the use of the water available for irrigation. A more systematic analysis of local conflicts about water (Böhmelt et al. 2014) revealed that most difficulties can be solved on a political or even personal level; few violent conflicts emerge around water. Intensified agriculture, urbanisation and climate change, however, are all factors that are likely to aggravate the necessity of good governance regarding water.

Transnationally, the same combination of market-based, hierarchical and cooperative settings can be identified. One example of an all-inclusive, long-term, cooperative setting, however, can be found in the Niger Basin (Cascao and Zeitoun 2010). There, a river basin organisation was created that addressed a wide range of water-related and development issues, including non-governmental stakeholders. In this and other international settings, many aspects of power asymmetries may weaken the cooperative aspects, including the nation's bargaining power (in the case of unequal neighbours such as India and Nepal) or the question of which nation is situated upstream and which downstream. "Struggles for equity, challenging power asymmetries and seeking sustainable access and allocation […] make up transboundary water politics" (Mirumachi 2015: 152).

When compared to Alpine grassland, pastoralism and Kibbutzim, cooperative water management is far more widespread. This may be perceived as an advantage, although it also makes it more difficult to identify common patterns that contribute to a thorough understanding of cooperative governance in the farming sector.

4.2   Linking with Consumers

Cooperative production is certainly strengthened by the desire to join forces in a strong and cooperative social setting. But this very concept of joining forces is not, of course, necessarily limited to agricultural production. Linking economic activities with cooperation is also attractive in other sectors of the economy. Moreover, it requires an institutional setting suitable for equitable, egalitarian and fair approaches. Over the last 200 years, it generally appeared that cooperatives represent the most appropriate institutional setting for such purposes. Indeed, it is only in recent decades that alternative institutional approaches have surfaced, some of which will be discussed below.

4.2.1   Cooperatives

Many British intellectuals recognised the failed attempts made during the early 19th century to organise a new cooperative means of production. Men, however, are able to learn from their mistakes. In 1833, 28 weavers managed to form the "Rochdale Society of Equitable Pioneers", a cooperative based in a small town north of Manchester, which served as a role model for cooperatives over a long period of time. The climax of this role model function was reached in 1937, when the International Cooperative Alliance adopted the original Rochdale Principles as the general standards for cooperatives. These principles were:

– The political power within cooperatives is equally distributed among their members. Although shareholding companies are usually controlled by the wealthiest members, in cooperatives each member contributes equally to the cooperative's capital and has one vote.

– Membership is open and voluntary. New members may not be discriminated against.

These two pillars summarise the most important characteristics of cooperatives. The Rochdale Society, which was principally formed to supply its members with a low-priced and reliable supply of food, also had additional regulations such as the obligation to pay in cash and the requirement for political and religious neutrality. These requirements, however, do not resonate to the same degree in today's cooperatives due to their being rather context-dependent. The original Rochdale Society ceased to exist in 1976, when it was merged with the neighbouring Oldham Cooperative.

In Rochdale, participants noticed that their purchasing power was considerably higher if they acted as a unit against other participants in the market. This was, during the 19th century, the strongest force behind the emerging cooperative movement. In Germany, for example, cattle traders charged inflated prices. If peasants needed to access credit in order to pay them, the interest rates were likewise far above reasonable levels. This encouraged Friedrich Wilhelm Raiffeisen (1818–1888) to found a cooperative bank run by and for farmers. Around the same time and just a few hundred miles to the east, Hermann Schulze-Delitzsch (1808–1883) founded a cooperative organisation that helped farmers to purchase inputs for less and sell their produce for more.

These stories have become long-lasting success stories. Yet, the decades of entrepreneurial history have, of course, caused changes. There changes include the Bavarian farming cooperative Baywa and the Austrian Raiffeisen Bank, in which the cooperatives have, at some point in time, decided to leave this particular institutional setting and convert into a shareholding company without the burdens associated with being a cooperative. There are even more examples of agricultural cooperatives gradually leaving the shrinking farming sector and offering services and products to a largely non-agricultural (although often still predominantly rural) population. However, there are also many cooperatives still serving farmers' interests, having farmers as members and farmers serving on the board.

Cooperatives have attained a role in various societies that justifies a closer look at this most significant institutional form of cooperative action. In fact, the resistance against cooperatives is the strongest in post-socialist countries such as Russia, where only 10% of the population are members of one or several cooperatives. In the West, the figures are usually higher, ranging from 17% of the population of the UK to 57% of the population of the USA.

The above overview has made it clear that the role of cooperatives in marketing agricultural factors and commodities exceeds the importance of the relatively few cooperatives that are still active in agricultural production. Chapter 2 has introduced the central position of family farming in today's agriculture. However, in Europe cooperatives organise more than 50% of factor purchases made by farmers and account for more than 60% of marketing activities concerning agricultural products.

Figure 4.1 offers an example of how many cooperative organisations work. As all farmers are supposed to participate and decide on issues in their cooperative, they

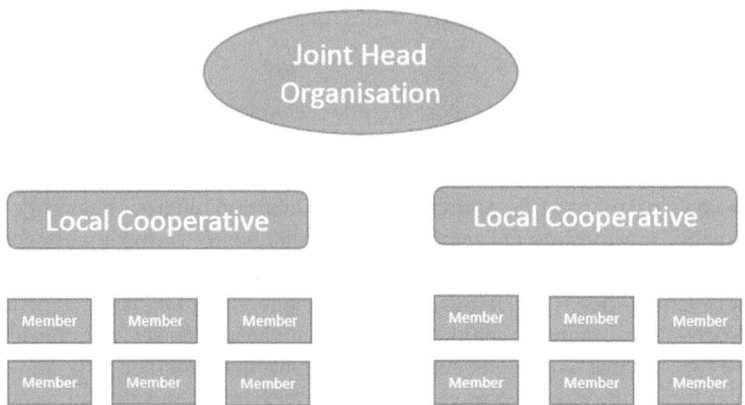

Fig. 4.1 Typical organisational structure of an agricultural cooperative

usually still work on the local or at least regional level. Nevertheless, they require national coordination, both in terms of negotiation power in the markets and gaining a voice in political issues. Therefore, local cooperatives often have a joint national organisation. On the EU level, there is even a General Committee for Agricultural Cooperation in the European Union, which includes fishery cooperatives. It is telling that this organisation shares an office with the interest organisation of European farmers.

While most cooperatives to date have preserved the basic building blocks of cooperative organisation, it has also appeared useful to some actors to dilute the organisational principles of cooperatives, mostly for pragmatic reasons such as better access to capital. Judge for yourself whether you consider Land O'Lakes, the biggest dairy in the USA, to represent such a dilution. They do both allow single farmers (7000 of their members) and local cooperatives (a further 1300 members) to join their membership ranks. Other developments are certainly more challenging to ideas of equality, including the so-called Wyoming cooperative model. Such organisations distinguish between patron members and investor members. While patron members are members of the cooperative as we know them, investor members simply bring capital into the organisation, meaning that they do not have voting rights.

New generation cooperatives (Harris et al. 1996) focus on delivery rights. Becoming a member of such an organisation means that you buy yourself the right (and the obligation) to deliver a defined amount of a certain agricultural commodity. The cooperative, in turn, is required to pay members a pre-specified price for the commodities delivered (usually a formula price based on spot market prices at a specified exchange, with additions or subtractions based on quality). The cooperative is also required to return any profits to members on a pre-specified schedule determined by the board of directors.

Garnevska et al. (2011) investigated a sample of Chinese cooperatives in order to determine which crucial success factors cooperatives should adhere to. They found that

– a stable institutional framework,
– an engaged general manager,
– support from the government and NGOs and
– engaged and understanding members

were the most important factors that distinguished successful cooperatives from the unsuccessful ones.

4.2.2 Fair Trade

Fair trade is a trading partnership based on dialogue, transparency, and respect that seeks greater equality in international trade. It contributes to sustainable development by offering better trading conditions to, and securing the rights of, marginalized producers and workers—especially in the South. Fair trade organizations (backed by consumers) are actively engaged in supporting producers, in awareness raising and in campaigning for chances in the rules and practices of international trade.

As this definition by DeCarlo (2007: 2) is endorsed by many major fair trade organisations, it seems well suited to briefly outline the concept that made the fair trade movement so big. As with cooperatives, it needed active groups and inspired individuals to turn the vision into practice, which happened more than 100 years later than for the cooperative movement. Generally speaking, two American NGOs with a Christian background, namely "Ten Thousand Villages" and SERRV, are considered to be the first organisations to have established trade with developing countries that paid a premium for, well, better development. That was in the 1940s.

The 1960s and 1970s saw the launch of the first shops specialising in solidary trade with developing countries. Yet, around the turn of the last century, the fair trade market enjoyed growth rates that other sectors could only dream of. Estimates suggest that 1.3 million farmers benefit from the fact that products worth 3.4 billion Euros in value are certified with a fair trade label.

Fair trade has become so successful due to combining two well-known patterns. One is that there is a steady demand for products in developed countries that can only be grown in southern countries, for example, coffee or bananas. The other pattern is that people in northern countries are often willing to transfer some of their wealth to the south, a transfer that has traditionally been accomplished via service projects. The combination of these patterns associated with fair trade takes advantage of the fact that people tend to feel more responsible for another person if they consume the products grown by that person. What economists refer to as a "warm glow" is achieved in a particularly sound manner if I can do something for the health and education of the farmer who grows the cocoa for my chocolate.

Smith (2009) collects and evaluates the criticisms directed towards fair trade, including the argument that support would be more effective if it was given directly to poor communities rather than taking a detour via traders and labelling organisations (as voiced by Henderson 2008). This argument neglects the fact that, should fair trade not exist, the buyers of fair trade products are unlikely to substitute their purchase premium for a donation. After a careful evaluation of these criticisms, Smith (2009: 34) contrasts "the largely unsubstantiated critiques of fair trade with the evidence which supports the use of the system as a way to build capacity that would otherwise not exist." Empirical findings show that fair trade increases both family income and credit availability and often results in an improvement in nutrition, health and education while also establishing economic capacity.

Nobody, however, doubts that the governance of fair trade could potentially be improved. When involving labelling organisations, there is always the danger of taken too much money out of consumers' payments for administrative purposes. Further, when paying prices above the equilibrium price, there is always the risk of encouraging production in locations where it is not efficient to produce. Attentive, idealistic and smart individuals are needed in the fair trade business to overcome such challenges.

4.2.3 Community-Supported Agriculture

The roots of community-supported agriculture lie in Japan during the 1960s. Consumers who were unhappy with mainstream retailers and an agricultural policy that gave contradictory signals to farmers and society started forming Teikei (Kondoh 2015). The Teikei were groups of farms and consumers that formed a relationship to govern the delivery of food, particularly milk, eggs and vegetables, that is, items with a low degree of processing. They were usually organic, which was almost the only well-organised distributional channel for organic food at that time.

One important element of Teikei groups is the voluntary work of their participants. Some Teikei groups allow members to buy themselves out of this obligation, although in general it is expected that members visit the farms in order to help them with the planting or harvesting work.

It is difficult to find reliable statistics concerning the extent of the Teikei movement, but some estimates (Kondoh 2015) suggest 25% of Japanese households are involved in one form or another. In fact, the peak of the Teikei concept was reached in the early 1980s. Paradoxically, the growing availability of organic food among Japanese retailers decreased the attractiveness of Teikei—now there were other ways to obtain organic products. Today's Teikei groups often suffer from the relatively high average age of their members. Young people often prefer to separate the buying of food from their working activities, and they are happy with what conventional retailers offer.

In most northern countries, the opposite is true. What the Japanese know as Teikei is an emerging social movement known as "community-supported agriculture" (CSA) in a growing area of the world. In the USA, 13,000 CSA farms deliver to their customers, of which the largest, "Farm Fresh to You" in California, serves 13,000 households. Likewise, Europe hosts at least a few hundred CSA networks, although it must be recognised that contradictory numbers are circulating.

While extremely similar to Teikei, it can be said that CSA rests on three main pillars that are interestingly homogeneous over the large number of countries in which such networks are emerging:

- One pillar is food security. It will be shown in detail below how many consumers have lost their confidence in the quality of the food that is offered by mainstream retailers. It strengthens trust considerably if consumers know the farm where their food comes from. Even more so, when they can discuss strategies concerning weed management or animal nutrition with "their" farmer.
- Most farms that participate in CSA produce organically. Even for those that do not, the attempt to choose sustainable production strategies can be considered another constitutional characteristic of CSA.
- Community-building is the third pillar of CSA. This includes the integration of voluntary work on the farm by group members as well as price setting. As a principle, the prices charged for the food are not market prices, but are instead supposed to reflect the true costs of production.

This latter point regarding price setting may sound good, but it faces conceptual challenges in relation to the farm manager's wage. A large share of production costs, particularly in small family farms, consists of the time the farm manager spends on production-related activities. Tegtmeier and Duffy (2005) highlight the frustration of many farmers participating in CSA who feel that their labour is not reimbursed in a fair way.

While CSA may appear to be an ideal means of bridging the large contemporary gap between farmers and consumers, it does face obstacles from the demand side. Empirical studies conducted on both sides of the Atlantic (Kato 2013; Maschkowski et al. 2017) show that food sovereignty is a serious issue. If the farm involved in your CSA project harvests beetroot, then beetroot will be delivered to your house, whether or not you might prefer beans. A long-lasting membership of a CSA project therefore presupposes a high degree of tolerance towards a broad range of different agricultural products.

4.3 Governing Sustainability

Few theoretical concepts have had such a large impact as the idea of sustainability. While there are discourses regarding a "sustainable industry" (Paton 2000) and even a "sustainable service sector" (Elekdag 2012), this affects agriculture more than any other sector. Agriculture, as with forestry, is strongly connected to natural processes that are dependent on the integrity of ecosystems. Yet, different to forestry, consumers let the final products of agriculture enter their body, which means that information concerning harmful substances is taken particularly seriously. It is therefore worthwhile to follow some of the processes for identifying a definition as well as the emergence of governance of sustainability, since the concept has certainly shaped agriculture as we know it today.

Sustainability is not fruitfully spread by giving orders, nor does it make sense to think of a "market" in which the proper understanding of sustainability is traded. The large and colourful debate around sustainability, in spite of major disputes and conflicts, is best understood as a story about cooperation, about a large part of society struggling to achieve a joint understanding of a new societal contract. It is therefore well placed within this chapter on cooperation.

4.3.1   The History of the Concept

After the historically unique phase of economic growth witnessed between 1800 and 1950, the second half of the 20th century was characterised by a growing awareness of both the negative side-effects and limits of growth. Rachel Carson (1907–1964) was a pioneer who played a prominent role in that process in the USA. After writing some books about maritime subjects, her final book, "Silent Spring", was concerned with the side-effects of pesticide applications. Following its appearance in 1962, the book was sold by the hundreds of thousands and it initiated the first broad debate about environmental issues.

Subsequently, the Club of Rome was founded in 1968 as a think tank. Among its first tasks was a modelling exercise. Founded by the Volkswagen Foundation, a computer simulation was used to see what would happen if a fixed stock of resources was faced with population growth and industrialisation. Different scenarios considered different extents of unknown reserves and different growth functions. As a result, petroleum reserves, for example, were projected to last between 20 and 93 years. The book in which these and other results were published in 1972, "The Limits to Growth", became the best-selling book ever on environmental issues. In the same year, the United Nations held its first assembly on environmental affairs.

Rachel Carson and the Club of Rome had brought environmental issues to the attention of a broad segment of society in industrialised countries, as had a number of obvious environmental issues that emerged during the 1970s and early 1980s, for example, acid rain and the destruction of the ozone layer. It also became increasingly obvious that these environmental issues should be considered within a larger socioeconomic frame, which should also take into account issues such as the widening gap between rich and poor countries. For then UN Secretary-General Javier Perez de Cuellar, 1983 marked the point at which to ask Norway's Prime Minister, Gro Harlem Brundtland, to create an organisation independent of the UN to focus on environmental and developmental problems and solutions. The World Commission on Environment and Development was hence founded in 1984.

The publication of the first volume of the Commission's report in 1987 marked the birth of the sustainability concept as we know it today. Indeed, the sentence

"Sustainable development is development that meets the needs of the present without compromising the ability of future generations to meet their own needs" will no doubt remain among the most cited sentences in human history.

The Brundtland Commission also established a second important conceptual foundation of sustainability. Sustainable development, they proposed, was based on the three pillars of economic growth, environmental protection and social equality. While some critics suggested the impossibility of sustainable growth (Daly and Townsend 1993), it is widely accepted today that sustainability rests on the three pillars of environmental, economic and social issues.

In 1987, that is, the same year in which the Brundtland Report went public, a rather elusive circle of social and natural scientists published both a book entitled "Ecological Economics" and a special issue of the journal "Economic Modeling" concerning the same issue. They attempted to take the specific characteristics of ecological systems into account, emphasising the irreversibility of many biological processes. Over the years, ecological economists have developed the concept of strong sustainability. This assumes that the existing stock of natural capital must be maintained, simply because the functions it performs cannot be duplicated by manufactured capital. While conventional environmental economists would be interested in the values of natural amenities, ecological economists would not really be concerned with them: We would just have the obligation to preserve nature as it is.

During the 1980s, the majority of farmers would not have considered that these debates would be of any relevance to them. They erred, however, at least in the medium term, since over the years, the consciousness regarding sustainability in all its different shades has been transmitted in different ways to agricultural practices. Many governments, particularly in Europe and North America, started to introduce incentives for farmers to help keep nature intact or at least to reduce the exploitation of natural resources. Linked even more strongly to the concept of sustainability, consumers were increasingly willing to pay extra for products that could evidence their sustainable production methods. The following sections are devoted to these initiatives and how they shaped the concept of sustainability, with an emphasis on the cooperative governance aspect.

4.3.2 Roundtables and the like

When consumers long for the taste of a strawberry, they will buy a strawberry. But what if millions or even billions of consumers want their food to be grown sustainably? What are they supposed to buy in order to have their demands met?

Attentive readers of Sect. 2.2.2 may recall the case of the Swiss Addax Bioenergy Company investing in large pivots of sugar cane in Sierra Leone. The bioethanol they were producing was to be sold as a fuel on the European market, although not because bioethanol would be a better or a cheaper fuel than petrol, but rather because substituting bioethanol for petrol would help to curb carbon emissions and protect finite resources. However, as bioethanol would only have its market share due to sustainability issues, it would be important to deliver some proof regarding the sustainability of the bioethanol's production process.

The European Commission is aware of this challenge and has approached it by requiring the suppliers of biofuels to subscribe to a certification scheme. The Commission offers 16 different international systems (plus the Austrian Agricultural Certification Scheme) to which suppliers can subscribe in order to prove their sustainability, including such colourful names as the "Biograce GHG calculation tool" and the "Red Tractor Farm Assurance Combinable Crops and Sugar Beet Scheme".

Addax Bioenergy chose the Roundtable of Sustainable Biomaterials (RSB), which, according to the description on their homepage, "offers trusted, credible tools and solutions for sustainability and biomaterials certification that mitigate business risk, fuel the bioeconomy, and contribute to the UN Sustainable Development Goals."

The RSB represents a good example of the large number of hybrid organisations to emerge with the objective of providing credibility. Among the members are producer organisations such as the United States' National Biodiesel Board, end users like Airbus, NGOs such as the World Wildlife Federation (WWF) and international bodies like the Food and Agriculture Organization (FAO).

These actors agreed on twelve "principles" that cover environmental subjects such as soil and water, but also socioeconomic issues like local food security and land rights. All these principles are broken down into a major number of minimum requirements that mostly read as rather specific and technical, for example, "relevant government authorities shall be included in the stakeholder process to ensure efficient streamlining of the process with legal requirements".

By using the example of Addax Bioenergy, it has already become obvious that the RSB, which is supported by the European Commission, is open to the certification of projects that some would accuse of unjustified "land grabbing". It is therefore logical that authors such as Goetz (2013) criticise the "insufficient protection of communities or the environment in weak regulatory settings and in view of intense commercial pressure on land" provided by the RSB.

This is a situation rather typical of the many attempts at sustainability definitions. The mixture of actors usually guarantees at least some degree of mixing of commercial and societal interests. Further, the resulting compromise is likely to prove unacceptable for least for some of the outside actors. This does not only apply to the many different roundtables within the realm of agricultural chains, but also to the systems provided in totally different actor constellations. For instance, it applies to sustainability strategies and labels used by major food companies such as Starbucks and Walmart, and it also applies to rather academic groups that issue their own sustainability assessment tools, for example, SMART (Schader 2016).

Globally speaking, the greatest top-down progress within the sustainability debate was probably made by the Brundtland Report in 1987. Since that time, many developments have taken place on the ground. The subsequent section will try to do justice to them.

4.3.3 Socioeconomic Sustainability Revisited

Typically, the environmental pillar of sustainability attracts the most attention, both inside and outside academia. In a textbook concerning the socioeconomics of agriculture, it is obvious that the opposite should be the case. There will be more convenient locations in which to debate the details of sustainable phosphorus or biodiversity management, but there are few more convenient locations to discuss issues related to socioeconomically sustainable agriculture.

It is extremely tricky, however, to appropriately deal with economic sustainability. The easiest part is the definition of sound indicators. Meul et al. (2008), for example, consider the production of value-added products, the efficient use of production factors and a low risk in agricultural production to be the core factors. Similarly, Ryan et al. (2016) suggest focusing on different productivity indicators. Such and similar suggestions may provide valuable clues with respect to the state of a farm (or any other enterprise). Yet, sustainability is also used to distinguish "good" farms from "bad" ones, both for marketing chains and for policy-makers. This is where economic sustainability indicators quickly reach their limits. Obviously, it makes no sense to only buy eggs from farms with high labour productivity or to restrict direct payments to farms above a certain income threshold. Economic sustainability on the farm level has few direct externalities, although a low level of sustainability may raise distributional concerns and might, in fact, be an argument for more rather than less public support.

This is different in the case of social sustainability. Social sustainability, to a great degree, refers to the links between the farm business and the outside world, which is a parallel to environmental sustainability. However, the flows in and out of the farm are not flows of nitrogen and pesticides, but rather flows of money as well as flows of appreciation and respect. The affected third parties are both farm workers and their relatives, in addition to neighbours and customers. Farms are a crucial component of the local social fabric and they should know about it.

It is a worthwhile exercise to choose 50 sustainability assessment tools in which social aspects are included in order to find out more about any emerging consensus concerning what social sustainability actually entails. When this exercise was completed for this book, it turned out that 24 of the tools were connected to a label, while 26 were assessment tools with a merely informative purpose. All the indicators were recorded and sometimes grouped into larger categories so as to see which subjects were the most frequent. It appeared that there were three main front runners:

- Thirty-two of the 50 frameworks contained safety hazards. Apparently, a farm is not socially sustainable if pesticides or machinery are handled in ways that provoke accidents or long-term damage. Indeed, the concept of sustainability is touched at its very core if incidents occur on farms in which some of the workers or even bystanders are harmed. Bennett (2013) reports 621 fatalities in US agriculture in 2010 and, while the figures from developing countries are less widely distributed, they are certainly not lower.

- Twenty-seven out of 50 frameworks include discrimination on their blacklist, mostly with respect to payments. While non-discrimination is an important component of just treatment, it is not as strongly related to the concept of sustainability as hazards. One could discriminate against minorities now, just as one could in the future. However, another core concept is the fulfilment of human needs. It is plausible that this will be difficult if parts of society are discriminated against.
- Another 27 frameworks included child labour on their indicator list. This nicely combines the two sides of the sustainability framework. Children require education (even if they are not aware of that in some cases), while our society needs well-educated children in order to maintain our standard of living in the future. In a sustainable agricultural system, children should therefore be spared from the obligation to work.

This shortlist does not pretend that more or less all frameworks concerning social sustainability would be the same. The SMART tool created by Schader (2016), for example, covers 40 different categories in the social realm, while the Sustainable Winegrowing New Zealand network only covers one category, namely training on the job for the winegrowers' employees. Moreover, there were several indicators only chosen by one of the 50 frameworks. Among them were the payment of bonuses to employees (in the label of the Food Alliance), the non-obligation for women employees to undergo pregnancy tests (in the Veriflora label) and the farm manager's tolerance for changes [in the MESMIS system constructed by Lopez-Ridaura et al. (2002)].

A question arises regarding whether the current governance of sustainability is sustainable in itself. From a radically libertarian standpoint, it is probably acceptable that different groups of people claim sustainability to have different attributes. It is the market's obligation to select concepts that are acceptable to consumers. This notion fails, however, in the presence of information asymmetries. It is reasonable to assume that a majority of wine buyers purchasing bottles with the logo of Sustainable Winegrowing New Zealand are not aware that the understanding of social sustainability is much more limited within this network than in most other labels. Would they consider themselves to be betrayed if they found out? Or, from a different perspective, are they willing to pay a higher price for a label emphasising bonus payments to employees, as in the Veriflora case?

In both the organic and fair trade cases, international actors have managed to establish a worldwide system featuring similar understandings of what the terms "organic" and "fair trade" actually refer to. However, in relation to sustainability in agricultural production, this is clearly not the case. The concept is apparently too broad for a global niche organisation (IFOAM in the case of organic production, WFTO in the case of fair trade) to establish such a definition. It might be possible to involve the International Standardization Organization (ISO), a body founded in 1947 to agree on and promote proprietary, industrial and commercial standards. While the core work of the ISO is rather technical, it might be time to create a consensus on what sustainable agricultural production is as well as to consider the realm of social issues.

Mann (2018) suggested one potential option: "The fulfilment of the individual subjective needs has to be aimed at in order to gain social sustainability. While the human rights set the bottom threshold and everything below them cannot be regarded as socially sustainable, sustainable development can be seen as development that increases the fulfilment of needs, hence moving up within the needs pyramid both within work and private life. To answer the initial question, a social farming system is then sustainable when the cultural institutional settings allow one to satisfy all needs or to improve the satisfaction of needs, both of physiological and social nature, and actors as well as institutions continuously recreate a system that allows future generations to do the same."

4.4 Concluding Thoughts on Cooperation

Institutional economists often emphasise that "old rules are good rules" (e.g. Kasper 2013). While this might be true for hierarchies and markets, the realm of cooperation may be particularly reliant on the firm establishment of cooperative structures. The long-term success of agricultural cooperatives and Alpine grassland corporations are two cases in point, while the many failed attempts to institutionalise cooperation in new ways are others.

However, new needs constantly emerge and, in the case of Western consumers, these needs often suggest interlinking more strongly with many of the poor persons responsible for growing the ingredients for food and drinks. A feeling of shared responsibility fosters the establishment of new and lasting organisational settings. Some consumers choose to link with farmers in their region through community-supported agriculture, while others buy fair trade items to help farmers in the global south. In any case, the need to buy sustainably is growing, and it is accompanied by both opportunities and threats. The key opportunity is to establish broadly accepted guidelines for the fairer organisation of production, which may become legal requirements one day. The main threat is the loss of credibility caused by freeloaders that use labels and symbols without actually delivering what they promise.

References

Bennett, C (2013) Death on the farm a grim reality of agriculture (Oct 24, 2017). http://www.westernfarmpress.com/blog/death-farm-grim-reality-agriculture

Böhmelt T, Bernauer T, Buhaug H, Gleditsch NP, Tribaldos T, Wischnath G (2014) Demand, supply, and restraint: determinants of domestic water conflict and cooperation. Glob Environ Change 29(2):337–348

Cascao AE, Zeitoun M (2010) Power, hegemony, and critical hydropolitics. In: Earle A, Jägerskog A, Öjendal J (eds) Transboundary water management—principles and practice. Earthscan, London

Daly HE, Townsend KN (1993) Valuing the earth: economy, ecology, ethics. MIT Press, Cambridge

DeCarlo J (2007) Fair trade beginner's guide. Oneworld Publications, Oxford

Elekdag S (2012) Social spending in Korea. IMF, New York

Engels F (1962) Die mark. In: Marx K, Engels F (eds) Werke, band 19. Dietz, Berlin

Garnevska E, Liu G, Shadbolt NM (2011) Factors for successful development of farmer cooperatives in Northwest China. Intern Food Agribus Manag Rev 14(4):69–84

Goetz A (2013) Private governance and land grabbing: the equator principles and the roundtable of sustainable biofuels. Globalizations 10(1):199–204

Haller T, Fokou G, Mbeyale G, Meroka P (2013) How fit turns into misfit and back: institutional transformations of pastoral commons in African floodplains. Ecol Soc 18(1):34

Harris A, Stefanson B, Fulton M (1996) New generation cooperatives and cooperative theory. J Coop 11(1):15–22

Henderson D (2008) Fair trade is counterproductive and unfair. Econ Aff 28(3):62–64

Kasper W (2013) Economic freedom and development. CCS, New Delhi

Kato Y (2013) Not just the price of food: challenges of an urban agriculture organization in engaging local residents. Sociol Inq 83(3):369–391

Konczacki ZA (2014) The economics of pastoralism: a case-study of Sub-Saharan Africa. Routledge, London

Kondoh K (2015) The alternative food movement in Japan: Challenges, limits, and resilience of the teikei system. Agric Hum Values 32(1):143–153

Lopez-Ridaura S, Masera O, Astier M (2002) Evaluating the sustainability of complex socio-environmental systems. the MESMIS framework. Ecol Ind 2(1–2):135–148

Mann S (2018) Conservation by innovation: what are the triggers for participation among Swiss farmers? Ecol Econ 146(2):10–16

Maschkowski G, Barth A, Köngeter A (2017) Solidarische Landwirtschaft – Austrittsgründe ehemaliger Mitglieder (October 6, 2017). http://ageconsearch.umn.edu/record/262180/files/Maschkowski_221.pdf

Merten S, Haller T (2008) Property rights, food security and child growth: dynamics of insecurity in the Kafue Flats of Zambia. Food Policy 33(5):434–443

Meul M, Van Passel S, Nevens F, Dessein J, Rogge E, Mulier A, Van Hauwermeiren A (2008) MOTIFS: a monitoring tool for integrated farm sustainability. Agron Sustain Dev 28(2):321–332

Mirumachi N (2015) Transboundary water politics in the developing world. Earthscan, London

Netting RM (1974) The system nobody knows: village irrigation in the Swiss Alps. In: Downing TE, Gibson M (eds) Irrigation's impact on society. University of Arizona Press, Tucson

Ostrom E (1990) Governing the commons. Cambridge University Press, Cambridge

Paton B (2000) Voluntary environmental initiatives and sustainable industry. Bus Strategy Environ 9(5):328–338

Russel R, Hannemann R, Getz S (2011) The transformation of the kibbutzim. Isr Stud 16(2):109–126

Ryan M, Hennessy T, Buckley C, Dillon EJ, Donnellan T, Hanrahan K, Moran B (2016) Developing farm-level sustainability indicators for Ireland using the Teagasc National Farm Survey. Ir J Agric Food Res 55(2):112–125

Schader C (2016) Nachhaltigkeit messen und bewerten. Ökologie & Landbau 2(2016):12–16

Smith AM (2009) Evaluating the criticisms of fair trade. Econ Aff 29(4):29–36

Tegtmeier E, Duffy M (2005) Community Supported Agriculture (CSA) in the Midwest United States: a regional characterization, 2005. Leopold Center for Sustainable Agriculture, Iowa State University, Ames, IA

White GF (1957) A perspective of river basin development. Law and contemporary problems 22(2):157–187

Yandle B (1974) Welfare programs and donor-recipient adjustments. Publ Financ Rev 2(3):276–291

Chapter 5
Agricultural System

The previous chapters have shown a lot of microstructures in the farming sector. Farmers and the other actors in the agricultural chain place themselves in hierarchies in which they dominate in some areas and are dominated in others. On various markets, they trade not only commodities but also labels, occupations and residences. Furthermore, they try to collaborate both in established and in new settings.

Generally, these hierarchical, market and cooperative dynamics apply to both Zambian pastoralists and Swiss Alpine farmers. Nevertheless, in reality, very few parallels exist between these two groups; their agricultural systems differ in almost every respect, also in socioeconomic terms. It is therefore worthwhile to look at agricultural systems as a whole. How do hierarchies, markets and cooperation add up to a real-life context?

Every system can be perceived from various angles, and this multitude will be addressed in this chapter. For example, a system can be perceived from a producer's or from a consumer's perspective, as presented in the following sections. Finally, the chapter will conclude with an 'objective', scientific perspective on agricultural systems.

5.1 Producer Perspectives

Philosophers, particularly if coming from a Marxist tradition, repeatedly emphasized that our social existence determines our consciousness. Thus, it will be useful to begin with a reflection on the social existence of farmers. What distinguishes traditions in farming from traditions in non-agricultural backgrounds? And how do the actual situations in current farming systems shift?

Affected parties usually have a different perception than parties not affected by a system. Thus, Sect. 5.1.2 will deal with empirical studies concerned with the self-understanding of farmers.

© The Author(s) 2018
S. Mann, *Socioeconomics of Agriculture*, SpringerBriefs in Economics,
https://doi.org/10.1007/978-3-319-74141-3_5

5.1.1 The Changing Environments of Farmers

We are coming from a past in which almost all humans were farmers. In northern countries, the 19th and 20th centuries were the time that reduced this vast majority to a tiny minority. In southern countries, this process started in the 20th century and is still ongoing.

This simple notion implies that farmers are usually the people in the region who do something similar to what their parents did, and in many cases they do not have to leave the region. As a logical consequence, farmers are more conservative and thus more reluctant toward change than non-farmers. This assumption has been confirmed for seven European countries (Baur et al. 2016), for Korea (Kyong-Dong 2003) and for the USA (Tickamyer 1983). It likely is true for almost every other region in the world.

Societal changes happen, regardless of farmers' resistance toward change. Some of these changes occur far away from most farmers' reality (such as changing attitudes toward homosexuality), but in other cases, the changes occur right on the farmer's doorstep. Such changes can be placed in four broad categories: agribusiness, consumers' demand, commodification and digitalization. Changes in agribusiness and changes among consumers are the two with the broadest validity. Both categories have been mentioned in a different context in previous chapters.

In agribusiness, for example, it has been mentioned that vertical integration, as a matter of hierarchy, increasingly restricts the decisions farmers can make, for example by dictating a lot of production parameters. However, vertical integration is not the only factor restricting farmers' choices. The concentration in the agribusiness sector has a similar effect. Market concentration has been taking place in most sectors of agribusiness, albeit to varying degrees. The degree of concentration in the pesticide business, tractor business and grain trading has always been high and is still growing. In the pesticide and tractor businesses, we still talk about six major market players each; the four largest grain traders cover by now over 90% of the market. This concentration leaves farmers with fewer and fewer choices which products to buy and to which buyer to sell.

The growing market share of hybrid seeds is another case in point subsumed under 'constrained choices' by Hendrickson and James (2005). For a long period, farmers had to make the choice whether to buy seed or to use part of their harvest as seed, which would generate slightly lower yields but save costs. The larger the share of seeds in hybrid varieties, the lower is this degree of freedom because hybrid seeds do not produce fertile seeds. In addition, a growing number of states also have implemented regulations forcing farmers to even pay fees to breeders if they use their own harvest's seeds. All of these developments may be justifiable in respect to innovation empowerment. However, none of them strengthens the farmer's leeway, and none will increase the pleasure of being a farmer.

Consumers' demand for sustainable practices, as discussed in the previous chapter, is a completely different driving force but may have a similar effect. Although conscious consumers can mostly be found in rich countries, their demand also covers the demand for bananas, coffee and other products that have to be grown in the global south. If farmers have the option to subscribe to labels that allow them to charge higher prices for their goods, the labels usually include additional restrictions: The animals have to be kept in special ways, the application of pesticides is only possible under certain conditions or not at all, or grassland has to be kept uncut until a certain date. The same applies to a growing number of agri-environmental schemes offered by governments mostly in Europe and North America. In some cases, production practices (e.g. keeping cattle outdoors in Switzerland) enable both a higher price through a label and additional public payments. The farmers will always happily accept the money. However, how are they affected by all the additional restrictions?

Burton and Wilson (2006) suppose that the restrictions change farmers' self-perception from productivism toward post-productivism. Although the paradigm of multifunctionality, describing that agricultural activities have all sorts of side-effects shaping society in many ways, has been around for some time, this paradigm may start to slowly alter the self-image of producers. Empirical studies that tested this hypothesis will be presented in the next subsection. Beforehand, the third and fourth categories of changes will briefly be touched. These types of changes are less universal but may considerably shape the development in some agrarian systems. Commodification and digitalization are very different drivers, but their impact should not be underestimated.

Commodification is a development that describes the increasing treatment of the inputs and outputs of agriculture as commodities. In the socioeconomic system of hierarchies, markets and cooperation, it could also be viewed as a development from cooperation to the market. Most scholars concerned with commodification in agriculture consider land as the most illustrative example, particularly in Africa (Bernstein 2007). De Janvry and LeVeen (1986) have been among the first to describe the integration of farming in the institutions of national and international markets, including the market for land.

Whereas the process of commodification can already fully be evaluated, it is certainly too early to fully judge the effects that digital technologies will have on agriculture. Only a few more or less speculative remarks can be made: It is unlikely that new technologies will narrow down the decisions farmers can make. To the contrary, their options of how to organize production should actually increase. However, intelligent technologies will decreasingly rely on human involvement. The farm manager would have the choice to run an automated farm on which the necessary work would be done without a lot of human involvement. If neighbours make the same decision, the size of the farm may have less and less to do with economies of scale. Drones and robots will not care whether they produce wheat on five farms of 20 ha or on one farm of 100 ha.

5.1.2 Empirical Results About Self-perception

If anything should have become clear over the course of this book, it is the strong heterogeneity of agriculture in different parts of the world. US entrepreneurs and Zambian pastoralists, for example, have very little in common. Under these circumstances, is it justified to say anything about self-perception of farmers? Perhaps self-perception of farmers differs too much within and between the various agricultural systems.

The multifaceted reality of farming systems barely allows finding a single term under which self-perception of today's farmers could be summarized. The best term would probably be 'productivist'. In general, farmers like to consider themselves as producers of food more than anything else.

This self-image applies, for example, to Kenya. As Waithaka et al. (2006) showed, Kenyan farmers, when depicting an 'ideal' farm, start dreaming about milk yields 10 times the actual level, or about corn yields 20 times as high as they are. Feed needs are underestimated, whereas animal density is rather overestimated. Such farms are ideal because they generate maximum food yield with minimum effort. The provision of anything else besides food, under these circumstances, is not really part of the picture. The self-image of being a productivist also applies to Ireland, where Howley et al. (2015) showed that farmers are willing to sacrifice additional income to avoid participating in agri-environmental or forestation schemes compromising their production potential.

However, it is crucial to acknowledge that farmers are as heterogeneous a group as people in most other professions. This heterogeneity is nicely illustrated in a study from the USA where Sulemana and James (2014) involved farmers in discourses about ethics. Farmers were confronted with scenarios such as applying pesticides under windy conditions or disposing pesticide containers without rinsing them. The farmers' attitudes differed widely, allowing the authors to categorize parts of their sample as conservationists, others as productivists.

A recent study from Switzerland (Mann 2018) raised the question of how 'green' even conservationists among farmers are. Farmers who subscribed to a public program on no-tillage and stated in a standardized survey conservation as a very important task for farmers would usually be regarded as being conservationists. However, in conversations with this group, this view was strongly challenged. The sequence below, for example, shows that these farmers, at least in some cases, still prioritize intensive production.

Farmer 3 (F3): Yes, and then I did it like that, and then I also worked outside the farm. Now I have arable production and pigs.
Interviewer (I): Yes, OK.
F3: And, er (...) right, since 03 I'm actually doing no-tillage (.)
I: OK
F3: When it actually was prescribed, due to run-off.
I: The district administration has prescribed it, right?

F3: Yeah, prescribed, they have recommended it, so to say, they actually recommended it.
I: Yes.
F3: And I thought, I would plough. Weed, right, problem, right?
I: Yes.
F3: And it is, of course, with the glyphosate that is, of course, (.) simple, I'm saying, no-tillage.
I: Yes, hmh.
F3: And I am always saying, if the glyphosate, the Roundup, we are actually only saying Roundup, if this goes away, ffff
I: Yes.
F3: Then I am seeing problems in arable production, right?

Although the farmer partly restricts himself of throwing in keywords like "weed" and "problem", the detailed text analysis in Mann (2018) shows how this farmer, as many others, leaves no doubt that he considers production as the primary objective and that he will need tools like glyphosate to maintain this objective.

Environmental conservation, however, is not the only requirement that can challenge the traditional farming perspective. The growing demand for animal welfare is a similar challenge to the traditional image of farmers as mere food producers. Te Velde et al. (2002) nicely showed how the self-images of farmers diverge from consumer perspectives in that respect. Farmers emphasize animals as a tool for production: Animals are supposed to serve for human nutrition, a process facilitated by farmers, and the farmers have no ethical issues with this view. Meat is and remains a necessary part of our diet. Consumers are much more ready to question the legitimacy of killing animals for nutritional purposes, even though their knowledge about the technicalities behind animal production is weak to non-existent. Subsequent studies (e.g. Franz et al. 2012) confirmed that animal behaviour will hardly become one of the main concerns of farmers, so that a gap between farmers' and consumers' perceptions is likely to remain.

The many new and extended activities of farmers, commonly known as farm diversification, are likewise challenging the traditional production-oriented view. Brandth and Haugen (2011) offered qualitative insights into this process. In interviews with Norwegian farming couples who are active in agritourism, they demonstrated how offering shelter, food, drinks and stories to tourists changes the self-understanding of farmers. These farmers fully perceived themselves as farmers but broadened the scope of being a farmer in a multifunctional direction.

Another branch of the literature follows the self-understanding of farmers as businessmen. Legally, managers of family farms are businessmen as much as any other self-employed entrepreneur. A Spanish survey (Gonzales and Benito 2001) showed that the majority of Spanish farmers consider themselves as workers ('trabajador') rather than businessmen ('empresario'). A follow-up study in Finland (Vesala and Vesala 2010) allowed its farming respondents to declare more than one identity. That way, only a quarter of farmers explicitly considered themselves as not being an entrepreneur. This comparison shows not so much the differences between agricultural reality in Finland and Spain, but more so the complexity of farming life, which can hardly be summarized in a single term.

Following the causalities of a considerable number of suicides among farmers in Australia and other countries, Bryant and Garnham (2015) described farm managers as 'fallen heroes'. They found a large gap between the romantic agrarian mythology of the 'Australian battler' who nurtures the population through hard work, struggle and self-sacrifice and the drought-stricken reality of volatile markets. This gap causes shame and despair ending lethally in some cases.

After all, a 'self-identity that is open to learning, difference and change' (Lankester 2012, 233) seems crucial for being prepared for the rapidly changing reality in the 21st century. Although this view generally applies to every professional group, the challenge is particularly large for the group of farmers, who over many centuries enjoyed a rather static perception of their duties in society.

5.2 Population Perspectives

Agricultural production, including agritourism or conservation, will always depend on the level of appreciation by the general population. In a way, farmers are on the safe side: As long as humans exist, they will be a very broad target group for food—nobody will be able to survive without food. Nevertheless, a strong dependence on how non-farmers perceive agriculture remains for various reasons.

The first dependency comes with shared localities for production. In developed countries, farmers have become a small minority, also in rural areas. Particularly for animal production, part of their viability will increasingly depend on which practices are tolerated in their neighbourhoods. On the consumption side, although it is true that consumers always have to eat, they are increasingly free to choose the origin of their food and the production system (integrated, organic, etc.). Finally, consumers are also political actors. A few of them are involved in designing new agricultural policy strategies, and many of them take part in elections where (among many other points) various agricultural policies are offered, of which some will generate more benefit to farmers than others.

It may be important to understand farmers' self-perception. However, to estimate how agriculture will develop in the future, it is arguably more important to understand the claims, hopes and (mis-)perceptions of the general population.

5.2.1 The Role as Residents

The issue of allowing or not allowing local production is only raised in regions in which farmers have become a small minority, and therefore is of minor relevance for most developing countries. For the rest, local residents likewise have few concerns with respect to arable farming or grassland production. The main field of conflict is animal production. One of the first studies in this field (Mann and Kögl 2003) was motivated by the experience of failed investments in northeast Germany. By that time, the animal density in the sparsely populated region was extremely low, so that politicians managed to attract some investors for large-scale (capacity 10,000 animals) pig fattening stables. However, only nine out of the 18 investments were realized, whereas the rest was prevented by local resistance. A combination of a survey and interviews with local mayors revealed some patterns important for acceptance, which differed considerably between villages with investments and villages without investment.

In villages without investment, the economic argument was the main driving factor. If residents believed a pig stable would generate new jobs and added value for the local economy, they would be in favour of it, but not if they did not believe in this economic effect. This economic concern faded as soon as the stables were built. From this time on, environmental factors became important. People who found the odour of pig farms problematic and who expected water pollution were now the ones against the investment.

Another factor that was crucial for the acceptance of the investment project and surfaced in the study was the social integration of the potential investor. Persons with a network in the village, perhaps by being sponsors of the local fire brigade, had much better chances to succeed with their investment project than foreigners with little relation to the target community. A related finding was validated when Soland et al. (2013) analysed the acceptance of Swiss biogas plants: Information offered to the local population increased trust and perceived benefits while reducing scepticism.

Gerlach and Spiller (2008) could not prove that decisions on new stables in rural areas would be based on negotiations. They considered farmers in a position so weak that the lack of clear legal guidelines would prevent many new investments. Given that the economic effects of animal production goes beyond the locality, their point may be an important one. Whereas radioactive waste depositories are often built without majority support of the local population, a society that wants to produce (or at least eat) meat should be able to define mechanisms that enable the investment in efficient production sites.

5.2.2 The Role as Consumers

It may not have been coincidental that the case of pig production was the starting point for our analysis of consumer attitudes. Many critical aspects—animal welfare, pollution and the health effects of meat consumption—culminate in this subsector. Two more publications on consumer attitudes toward pig production, albeit from rather differing cultures, can help understand the dynamics. de Barcellos et al. (2013) asked Chinese consumers about their demands regarding pig production to generate clusters with relatively homogeneous claims. They finally described three clusters. One cluster of consumers focused on food security, largely neglecting quality aspects.

This was the cluster with the lowest consumption but with the strongest preference for Chinese races. The second cluster was labelled as 'indifferent' by the authors. These consumers preferred medium-sized farms (i.e. around 400 animals) and placed the largest emphasis on taste, being the most active consumers of pork. The third cluster favoured large-scale industrial production, preferring lean meat imported from Britain.

Weible et al. (2016) followed a very similar objective when approaching German consumers. However, their methodology included a factor analysis beforehand to identify groups of variables describing relevant attitudinal dimensions. These dimensions consisted in a generally critical approach toward pig production, a critical attitude toward farmers, the acceptance of the existent system and a critical attitude toward other persons' behaviour. The subsequent cluster analysis resulted in the three groups 'opponents', 'moderates' and 'the tolerating'; the opponents, for example, scored high in critical attitudes toward pig production, toward farmers and toward other persons' attitudes but scored low in acceptance of the current system.

By comparing the two studies, one can certainly draw conclusions regarding different attitudes between German and Chinese consumers. However, the two project teams had differing foci. It is no coincidence that the Chinese questionnaires focused on the aspects of food safety and quality, whereas the German survey focused on animal welfare and the share of meat in people's diets. Both projects caused massive blind spots in their results due to the limited range of questions. This limitation was probably necessary to prevent huge, time-consuming questionnaires, which nobody would want to fill in. However, it remains central to keep the limitations of such focused research in mind.

Comparative research offers the advantage that the same research design can be applied to different systems, as done in a comparison of internet discourses on agriculture led by German or Swiss non-agricultural citizens (Mann 2015). Although the languages overlapped, the internet message boards and newsrooms were sufficiently separated to make this comparison. Both the quantitative and the qualitative part of the study generated the same result: In Switzerland, the discussions focused on the things that could be improved around agriculture. Participants discussed best practice or weather events affecting agriculture. In Germany, discourses had a different focus. Agriculture was more often than not considered as a dangerous black box. Food not produced by organic farmers was thought likely to be harmful, and agricultural production was viewed as harming the environment and animals. Although Swiss discussants also made critical remarks, they clearly moved *within* the system rather than feeling threatened *by* the system. The different agricultural structures could be one clue to understand the differences. The 20-ha farms in Switzerland may still allow at least indirect ties to farmers, whereas this contact may have vanished for a large majority in Germany, where average farms are three times as large.

Certain attributes, of course, can improve the perception of food quality. Chapter 3 repeatedly mentioned that organic production over a long period has done an excellent job to improve this level of trust. However, observations over time indicated that labels advertising production systems, even including organic production, lose importance over time, while the roles of local production (Moser et al. 2011) and quality grading (Mann and Erdin 2016) increase.

The issue of local production deserves a bit more attention due to the problem of low production site tolerance by the local population mentioned in the preceding subsection. Among a series of studies indicating a general preference among consumers for food produced in their country, Lobb and Mazzocchi (2007) stand out for the width of their approach. They conducted 2725 face-to-face interviews in five European countries and found preferences for domestic production in each. While trying to decompose the preference for local production among US consumers, Darby et al. (2008) found that production 'within the state' was similarly important as production 'nearby'. All this shows, of course, the contradiction between our two roles as residents and consumers with respect to animal-based food: As consumers, we want it to be produced close to where we live. As residents, however, we do not want animal farms too close to our homes.

5.2.3   The Role as Citizens

In addition to being residents and consumers, we are citizens. In this role, we are supposed to develop convictions that influence our voting behaviour. In most cases, it is not a party's stance on agricultural policy that makes us vote for it. Nevertheless, our political attitudes with respect to agriculture are an important component of the socioeconomic system of agriculture.

It may well be that the role as citizen contradicts the role as consumer. The German 'Agrarwende' at the beginning of the 21st century is a case in point. At that time, the share of organic products on the market was still around 2%. Then a member of the Green Party became minister of agriculture and announced strong financial support for organic farmers. A survey (Mann and Mante 2003) showed that this political strategy found broad support, even among conservative voters. Apparently, people who were not willing to spend money for buying organic food still seemed to be willing to spend tax money to promote organic food.

Usually, however, our attitudes toward agricultural policy are well embedded in our general belief system. This was shown by a study on the admission of genetically modified crops in Switzerland (Schläpfer 2008), in which voting behaviour toward a five-year moratorium of genetically modified crops was explained by survey results. The predictive power of sociodemographic characteristics such as education, age or gender was very weak. Related attitudes did a much better job. Respondents concerned about the freedom of research were consistently against such a moratorium, whereas respondents concerned about the health and environmental effects of genetically modified organisms were in favour.

Mittenzwei et al. (2016) attempted to explain the origin of such attitudes and found support for the hot cognition theory. This theory suggests that our culture, the milieu we come from, shapes our attitudes. Knowledge is then only used to find supportive arguments for this attitude, not to change it. Accordingly, the authors found that the

level of knowledge we have about agriculture does little to alter our attitudes toward
agricultural policy. Growing up in a rural area and having farmers among your friends
are much stronger predictors of your evaluation of farming policies than your related
knowledge.

Results by Aerni et al. (2009) suggested that not only do attitudes influence
agricultural policies, but also do agricultural policies influence attitudes. They
showed that stakeholders in New Zealand, where farmers operate in a free market,
pursued a more innovative approach toward the concept of sustainable agriculture
than stakeholders in Switzerland, where the state provides conservative political conditions and where maintaining the status quo has a high priority among stakeholders.

5.3 Varieties of Capitalist Agriculture

After presenting perspectives through the eyes of others, it may now be overdue to
return to the diversity approach as presented in Sect. 1.3. Although authors writing
about the varieties of capitalism have defined emerging forms of market economies,
the analysis of the varieties of capitalist agriculture is still due. What should be the
differences in the approach?

The agricultural sector strongly depends on land; furthermore, it is older than the
industrial and service sectors and it targets basic needs. Thus, some of the character-
izing variables of agriculture will probably be different from those of other sectors.
This claim becomes clearer when we apply the five core blocks of variables proposed
by Amable (2003) for the farming sector.

One case in point is the wage labour nexus used to characterize varieties of
capitalism. In developing and developed countries, family farming is the dominant
form of production. This implies that wage-dependent labour has a far lower impact
than in other sectors. The organization of financial systems may also have some
importance for the organization of farming, but financial services are a sector of
their own, distinct from agriculture. The main tools of social policy are primarily of
relevance for urban areas (Todaro and Stilkind 1981; Mann 2005), so that the focus
of these policy instruments is not appropriate for an understanding of agriculture.
Education may be more relevant for agriculture than the indicators mentioned above,
but probably less so than for other sectors. Hence, four of the five blocks used by
Amable (2003) to characterize the diversity of capitalism are of very limited use for
describing the agricultural sector.

From Amable's set of choices, the product-market-regulation variables may be
the most relevant ones for the farming sector. The level of protection in agriculture
is markedly higher than in the other two sectors (Josling 2000; Morley and Piñeiro
2007; Matsumura 2008). Of 1 Dollar earned, sometimes more than 50 Cents come
from tax money, mostly through direct transfers to farmers and market support. This
fact, of course, has grave consequences for the entire sector and individual farming
strategies.

Agricultural economists widely acknowledge that governments differ greatly in their support of agriculture. Brunstad et al. (1999), for example, recall Norway, Switzerland, Iceland, Japan and Finland as the 'biggest spenders of OECD' (p. 541). This finding is regarded either as welfare-destroying misbehaviour (Tyers and Anderson 1988; Hertel and Keeney 2006) or as a conscious strategy or view labelled as multifunctionality (Paarberg et al. 2002; Wüstemann et al. 2008). Multifunctionality emphasizes the importance of environmental amenities provided by farmers in addition to mere food production.

Potter and Tilzey (2005) identified three types of discourses in agriculture: neoliberalism, where most interventions in the sector are viewed as being welfare decreasing; neomercantilism, where national sectors attempt to protect themselves from foreign export interests; and multifunctionality, where public intervention is considered as internalizing the external effects of agriculture. However, Mann (2016) claimed that only neoliberalism and multifunctionality, due to a strong welfare–economic theoretical backbone, would qualify as paradigms.

There is thus a strong normative discourse among agricultural experts about the 'right' strategy for their sector. The lessons to be learned from the debate around the diversity of capitalism—complementarities that allow for certain characteristics of a society and not for others—have not yet been learned in the farming sector. It is therefore worthwhile, before closing this book, to leave aside the normative debate and empirically analyse the existing diversity of capitalist agricultural systems. Scholars concerned with empirical work on the diversity of capitalism have generally used cluster analysis to identify similar patterns among countries (Amable 2003; Farkas 2011; Schneider and Paunescu 2012). There is no reason to change this approach when shifting attention from the national to the sectoral level. However, there is a reason to start afresh by identifying appropriate variables for our purpose.

5.3.1 Selection of Variables

As mentioned in the previous section, governmental support plays a significant role in shaping the agricultural sector. In particular, tariffs for food imports and direct transfer payments to farmers are instruments which are still broadly applied to protect domestic production. The producer support estimate (PSE) by the OECD (2016) has been for many years a widely accepted measure used to quantify the support given to the farming sector. Although the PSE is measured in absolute money terms, it becomes more meaningful if set into relation with gross farm revenues. This number, the percentage PSE, describes how many cents of one dollar a farmer owes to the state's agricultural policy.

In some countries, not only producers but also consumers benefit from generous public policies making food more affordable. Producer and consumer support estimates add up to the total support estimate (Tangermann 2004; OECD 2016). If set into relation with the country's gross domestic product, the ratio gives a useful impression of what share of national wealth is used to keep farmers and food consumers happy—or, if negative, how the food sector is used to fund other parts of the economy through taxation, for which Ukraine would be an example.

The size of farms also shapes the agricultural system. Although many possible ways exist to statistically measure farm size (Mann et al. 2013), a global comparison is well advised to focus on acreage. It is obvious that an average Chinese farm with 0.7 ha must be organized along different lines than an average Australian farm with 3200 ha. Lowder et al. (2016) provided an excellent overview of the frontiers of knowledge regarding worldwide farm sizes. Admittedly, for some countries with a strongly bifurcated agricultural structure, the information provided by this variable is of only limited use. For example, Russia and South Africa have two coexisting agricultural systems in their countries: big commercial farms and a large number of smallholders (Greenberg 2010; Lerman and Sedik 2013). The average for these cases is therefore of little importance. However, we accepted this weakness in light of the precious information the variable provides in most other cases. Consequently, we kept South Africa and Russia in the sample.

Another indicator would be trade balances. Agriculture in countries where food is the main export item will have a different status than agriculture in countries where it mainly competes with imports. Most self-sufficiency measures compare calories produced with calories consumed (Pinstrup-Andersen 2009). The Food and Agriculture Organization uses also monetary figures. This value may give a more balanced picture of the trade balance because it considers the value of the traded goods, so we used it for our analysis.

Agriculture is a sector with major environmental impacts, accounting for 9% of worldwide greenhouse gas emissions and being the most important emitter of methane and nitrous oxide (Sensi 2016). The resource efficiency of agriculture has become a central concept for scientists (de Wit 1992; Hayashi 2000; Keating et al. 2010; Altieri et al. 2012) and policy makers. As the Food and Agriculture Organization has collected and published estimates on emissions of nitrous oxide and methane per country, it is useful to set these emissions into relation with the agricultural outputs of the countries concerned, as a rough estimator of environmental resource efficiency.

Last but not least, we included food expenditure per head as a clustering variable. Although food expenditure is usually considered as a proxy for food security (Esturk and Oren 2014) or income (Oyekale and Adesanya 2012) in poorer countries, it does not lose its relevance in wealthier regions. The costs borne by households to feed themselves are a good descriptor of the interplay between food prices on the one hand and purchasing power on the other. The amount spent on food also reflects quality components that are difficult to operationalize and many factors from the agri-food chain that are likewise difficult to grasp.

Table 5.1 Variables used for the description of varieties of agriculture

Variable	PSE (%)	TSE (%)	Farm size (ha)	Self-sufficiency (%)	CH$_4$	N$_2$O	Expenditure (US$)
Explanation	Percentage producer support estimate	Percentage total support estimate of gross domestic product	Average farm size in ha	Value agricultural products consumed as percentage of value agricultural products produced	CO$_2$ equivalents of methane divided by food production	CO$_2$ equivalents of nitrous oxide divided by food production	Food expenditure per head in US$
Mean	18	0.88	284	139	0.87	0.69	562
Minimum	−7	−3.05	0.7	84	0.06	0.12	245
Maximum	62	4.57	3243	526	1.88	1.30	1117

5.3.2 Processing of Variables

The variables described in the previous subsection are summarized in Table 5.1. The question for which countries these variables should be collected and processed is answered through data availability and conception issues. On the latter issue, most empirical studies restrict themselves to wealthier countries, because the 'varieties' otherwise would often just distinguish poorer from wealthier countries, as Solga (2014) explains.

As another distinction from clustering exercises on general economic characteristics, it does not make sense to treat European countries separately. For more than 50 years, the European Union (EU) has enjoyed a common agricultural policy, so that important characteristics are no longer nation specific, particularly not the degree to which agriculture is subsidized. Therefore, the EU was treated as an entity in the analysis.

K-means (Steinhaus 1956; Jain 2010) as the most established algorithm of cluster analysis was used in Stata. The average farm size was eventually transformed into a logarithmic scale to avoid a too-powerful influence on the outcome. After various attempts, we decided that dividing participating countries into three groups would generate the highest explanatory value.

Table 5.2 Results of the cluster analysis

Cluster	PSE (%)	TSE (%)	Farm size (ha)	Self-sufficiency (%)	CH$_4$	N$_2$O	Expenditure (US$)
1	54	1.2	4.7	84	0.41	0.32	832
2	16	1.5	17.3	120	0.81	0.69	535
3	1	−0.2	127.7	200	1.21	0.91	451

5.3.3 Results

The three clusters are summarized in Table 5.2. Cluster 1 is the smallest of the three, containing Japan, South Korea, Norway and Switzerland. As an average, more than every second dollar earned in these countries is politically induced. This public support apparently comes to farmers by way of direct payments, rather than through artificially high food prices, as can be seen from the moderate total support estimate. As this cluster contains Korea and Japan, two countries with average farm sizes of just over one hectare, it is hardly surprising that this cluster has the smallest farm size. It is the only cluster with net food imports. The differences between the three clusters concerning environmental performance are considerable. It is obvious that Cluster 1 with its protective and small-structured approach produces much lower emissions per unit of production than the other clusters. Per capita expenditures on food are considerably higher than in other countries, and Switzerland (with 1100 US$ per person and year) holds the top place.

On the other end of the global spectrum, Cluster 3 unites countries that are much more directed toward free markets. It contains New Zealand, Australia, Brazil, Chile, Ukraine, Vietnam and South Africa, with average farm sizes of over 100 ha. These countries come closest to free markets of the global community. As an average, they largely abstain from subsidizing either farmers or consumers, although some participants such as Ukraine (total support estimate = −3.05%) are effectively subsidizing food prices instead of increasing them. Fifty per cent of the food output in these countries is exported to other countries. It seems that the price for this expansive strategy is high emissions per unit of food produced.

Cluster 2, the largest cluster, contains Turkey, Russia, Kazakhstan, Israel, Columbia, China, Canada, the USA and the EU. Although food expenditures per capita are considerably lower than for Cluster 1, Cluster 2 is the group with the highest taxation on food products. All other measures are situated between the two other clusters. There seems to be a broad middle course between a strong export strategy with large farms and cheap food at the expense of the environment and a greener strategy based on small farms, generous subsidies and food imports.

5.3.4 Discussion and Conclusion

The clustering on a sectoral level (i.e. within agriculture) revealed some remarkable results, particularly if compared with cluster results on the macro level as obtained, for example, by Amable (2003). The diversity of capitalism becomes even more diverse when broken down on a sectoral level.

An initial finding is that the clusters on the meso level, at least in the case of agriculture, diverge strongly from the results on the macro level. Canada and the USA, for example, share a cluster in both cases, but on a macro level they join Australia, which in the agricultural clustering (on the meso level) is in a different grouping. In the agricultural analysis, Switzerland is in company with Japan and South Korea, whereas the latter two form a cluster of their own in Amable's (2003) macro-level study.

The results reveal a peculiarity of the agricultural clusters. Compared with clusters from macro-level analyses, the sectoral clusters reveal far fewer geographical patterns. Cluster 1, for example, may be shaped on the one hand by the historical experience that self-sufficiency is a worthwhile goal, and on the other hand by climatic and topographic factors making self-sufficiency difficult. However, Norway and South Korea, for example, have almost no commonality beyond that, neither culturally nor geographically.

It is certainly worthwhile to reflect on both the causes and the impact of these differences. Some scholars have already linked different attitudes to different policies. Aerni (2009) showed that citizens in New Zealand consider agriculture in the context of agricultural competitiveness, whereas Swiss citizens watch new technologies with scepticism when it comes to sustainability aspects. This example indicates that different attitudes among voters might cause different varieties of capitalist agriculture; other branches of the literature also named history as a crucial factor. Spoerer (2015) nicely showed how disadvantaged farmers in the EU managed to make the moral case for a welfare policy in favour of the farming sector. In Australia, where agriculture does not have the traditional face but is rather considered as another entrepreneurial activity, this would not have been possible.

The three clusters provide some added value for the intra-agricultural discourse. For example, the common assumption that Japan, Switzerland and Norway are protective in terms of agricultural trade and pursue the model of multifunctional agriculture is much more often put into a context with the EU than with South Korea (e.g. Brunstad et al. 1999). Thus, the exercise of using sectoral variables for clustering reveals some new patterns.

The results on the sectoral level may be slightly less interesting than those on the macro level, where multi-dimensionality is one of the greatest assets. Finally, the three agricultural clusters can be placed on a rather one-dimensional scale. On one end of this scale, we observe an import-dependent agriculture that enjoys ample subsidies and produces high-priced food but has relatively low emissions per output. On the other end of this scale, a strong and export-oriented sector is doing well without state involvement, while causing environmental pollution. Most countries are between these two extremes, feeding themselves with some support for the farming community. This finding indicates, as a worldwide pattern, that societies are willing to transfer resources to farmers to substitute imports. When enough food is available for the population, the rationale for this transfer is apparently lost. The connection to the level of pollution certainly deserves increased future attention.

Still, the main advantage identified in the 'varieties of capitalism' debate certainly also holds for agriculture. The concept teaches us to emphasize complementarities rather than (sometimes artificial) welfare effects. Thus, worldwide agriculture can be seen as a colourful and rich composition of various fruitful systems.

5.4   Concluding Thoughts on Agricultural Systems

Many socioeconomic systems of organizing food production have been described in this book, both on the micro and on the meso level. Their diversity indicates that agricultural systems may have the potential of successful self-regulation. Based on this conclusion, we could interpret the developments of the last years and decades as follows:

- Liberalization processes in many countries could have been the response of unnecessary inefficiencies in government regulations and a move toward more affordable food.
- The demand for blooming meadows, butterflies and the like in many countries has been answered by agri-environmental programs.
- Mistrust about the side-effects of modern production methods has been met both by growing organic markets and by creating trust through community-supported agriculture.
- A perception of unjust resource allocations between northern consumers and southern producers has led to the fair trade movement.

This interpretation is not a claim that we live in the best of all possible agrarian worlds. The sectoral analyses on the micro and meso levels certainly revealed problematic issues for which solutions still need to be found:

- Many people care for animals today. This care is certainly reflected in the lives of cats and dogs, but hardly reflected in the lives of most pigs, chicken and cows. Animal welfare today is neither conceptually fully understood nor realized to a degree that would suffice for a large part of the population.
- Agriculture contributes to 15–20% of climate change, more so through animal production than through crop production, and little is done to reduce this contribution.
- The share of starving people on our planet has been reduced, but the share of obese people is strongly on the rise. Although the joy of overeating may outweigh the 'cost' of a belly, it is likely that we are actually moving away from the social optimum in this respect.

This book was extremely brief on these three aspects, simply due to the lack of promising solutions. However, all three issues could be tackled by reducing meat consumption. There is ample room for agricultural (and other) researchers to develop strategies to overcome these and other challenges.

A socioeconomic perspective, however, will certainly help to tackle these and other contemporary problems of agriculture, as it neither neglects the objective scarcities in the system, nor the cultural setting in which interaction and decisions occur.

References

Aerni P (2009) What is sustainable agriculture? Empirical evidence of diverging views in Switzerland. Ecol Econ 68(6):1872–1882

Aerni P, Rae A, Lehmann B (2009) Nostalgia versus Pragmatism? How attitudes and interests shape the term sustainable agriculture in Switzerland and New Zealand. Food Policy 34(2):227–235

Altieri MA, Funes-Monzote FR, Petersen P (2012) Agroecologically efficient agricultural systems for smallholder farmers: contributions to food sovereignty. Agron Sustain Dev 32(1):1–13

Amable B (2003) The diversity of modern capitalism. Oxford University Press, Oxford

Baur I, Dobricki M, Lips M (2016) The basic motivational drivers of northern and central European farmers. J Rural Stud 46(2):93–101

Bernstein H (2007) Capitalism and moral economy: land questions in Sub-Saharan Africa. http://citeseerx.ist.psu.edu/viewdoc/download?doi=10.1.1.466.6137&rep=rep1&type=pdf (30 Oct 2017)

Brandth B, Haugen MS (2011) Farm diversification into tourism—implications for social identity? J Rural Stud 27(1):35–44

Brunstad RJ, Gaarland I, Vårdal E (1999) Agricultural production and the optimal level of landscape preservation. Land Econ 75(4):538–546

Bryant L, Garnham B (2015) The fallen hero: masculinity, shame and farmer suicide in Australia. Gend Place Cult 22(1):67–84

Burton RJF, Wilson GA (2006) Injecting social psychology theory into conceptualisations of agricultural agency: towards a post-productivist farmer self-identity? J Rural Stud 22(1):95–116

Darby K, Batte MT, Ernst S, Roe B (2008) Decomposing local: a conjoint analysis of locally produced foods. Am J Agr Econ 90(2):476–486

de Barcellos DM, Grunert KG, Zhou Y, Verbeke W, Perez-Curto FJA, Krystallis A (2013) Consumer attitudes to different pig production systems: a study from mainland China. Agric Hum Values 30(3):443–455

De Janvry A, LeVeen EP (1986) Historical forces that have shaped world agriculture: a structural perspective. In: Dahlberg KA (ed) New directions for agriculture and agricultural research. Rowman & Littlefield, New York

de Wit CT (1992) Resource use efficiency in agriculture. Agric Syst 40(1–3):125–151

Esturk O, Oren MN (2014) Impact of household socio-economic factors on food security: case of Adana. Pak J Nutr 13(1):1–6

Farkas B (2011) The Central and Eastern European model of capitalism. Post-Communist Economies 23(1):15–34

Franz A, Deimel I, Spiller A (2012) Concerns about animal welfare: a cluster analysis of German pig farmers. British Food J 114(10):1445–1462

Gerlach S, Spiller A (2008) Stallbaukonflikte in Nicht-Veredlungsregionen: Welche Faktoren beeinflussen den Konfliktverlauf? In Spiller A, Schulze B (eds) Zukunftsperspektiven der Fleischwirtschaft. Universitätsverlag, Göttingen

Gonzales JJ, Benito CG (2001) Profession and identity. The case of family farming in Spain. Sociologia Ruralis 41(3):343–357

Greenberg S (2010) Status report on land and agricultural policy in South Africa. University of Western Cape, Stellenbosch

Hayashi K (2000) Multicriteria analysis for agricultural resource management: a critical survey and future perspectives. Eur J Oper Res 122(2):486–500

Hendrickson MK, James HS (2005) The ethics of constrained choice: how the industrialization of agriculture impacts farming and farming behavior. J Agric Environ Ethics 18(3):269–291

Hertel TW, Keeney R (2006) What is at stake: the relative importance of import barriers, export subsidies and domestic support. In: Anderson K, Martin W (eds) Agricultural trade reform and the Doha Development Agenda. World Bank, Washington

Howley P, Buckley C, Donoghue CO, Ryan M (2015) Explaining the economic 'irrationality' of farmers' land use behaviour: the role of productivist attitudes and non-pecuniary benefits. Ecol Econ 109(2):186–193

Jain AK (2010) Data clustering: 50 years beyond k-means. Pattern Recogn Lett 31(4):651–666

Josling T (2000) New agricultural negotiations: an overflowing agenda. Fed Reserve Bank St. Louis Rev 82(4):53–76

Keating BA, Carberry PS, Bindraban PS, Asseng S, Meinke H, Dixon J (2010) Eco-efficient agriculture: concepts, challenges and opportunities. Crop Sci 50(S1):S109–S119

Kyong-Dong K (2003) Presidential election and social change in South Korea. Dev Soc 32(2):293–314

Lankester A (2012) Self-perceived roles in life and achieving sustainability on family farms in North-eastern Australia. Aust Geogr 43(3):233–251

Lerman Z, Sedik D (2013) Russian agriculture and transition. In: Alexeev M, Weber S (eds) The Oxford handbook of the Russian economy. Oxford University Press, Oxford

Lobb AE, Mazzocchi M (2007) Domestically produced food: consumer perceptions of origin, safety and the issue of trust. Acta Agric Scand Sect C Food Econ 4(1):3–12

Lowder SK, Skoet J, Raney T (2016) The number, size, and distribution of farms, smallholder farms, and family farms worldwide. World Dev 87(4):16–29

Mann S (2005) Implicit social policy in agriculture. Soc Policy Soc 4(3):271–281

Mann S (2015) Web-based discourse on agriculture among the general public—comparing Germany and Switzerland. British Food J 117(1):388–399

Mann S (2016) The two competing paradigms of liberalism and multifunctionality in agriculture—a utilitarian perspective. AgroLife Sci J 5(1):121–126

Mann S (2018) Conservation by innovation: what are the triggers for participation among Swiss farmers? Ecol Econ (forthcoming)

Mann S, Erdin D (2016) Grades or labels? Beef prices and quality. Int J Qual Reliab Manag 33(9):1406–1412

Mann S, Kögl H (2003) On the acceptance of animal production in rural communities. Land Use Policy 20(3):243–252

Mann S, Mante J (2003) Die Agrarwende im Spiegel der Bevölkerung. Berichte über Landwirtschaft 81(2):302–315

Mann S, Mittenzwei K, Hasselmann F (2013) The importance of succession on business growth: a case study of family farms in Switzerland and Norway. Yearb Socioeconomics Agric 2013:103–131

Matsumura A (2008) Agricultural trade liberalization and human rights: economic analysis for poverty reduction in LDCs—a survey. Forum Publ Policy, Tokyo

Mittenzwei K, Mann S, Refsgaard K, Kvakkestad V (2016) Hot cognition in agricultural-policy preferences in Norway? Agric Hum Values 33(1):61–71

Morley S, Piñeiro V (2007) The impact of CAFTA on growth and poverty in four countries in Central America: evidence from a CGE analysis. IFPRI, Washington

Moser R, Raffaeli R, Thilmany McFadden D (2011) Consumer preferences for fruit and vegetables with credence-based attributes: a review. Int Food Agribusiness Rev 14(2):121–142

OECD (2016) Agricultural policy monitoring and evaluation 2016. OECD, Paris

Oyekale AS, Adesanya YA (2012) Climate change and urban children's health: a case study of Ibadan. Life Sci J 9(3):123–155

Paarberg PL, Bredahl M, Lee JG (2002) Multifunctionality and agricultural trade negotations. Appl Econ Perspect Policy 24(2):322–335

Pinstrup-Andersen P (2009) Food security: definition and measurement. Food Secur 1(1):5–7

Potter C, Tilzey M (2005) Agricultural policy discourses in the European Post-Fordist tradition: neoliberalism, neomercantilism and multifunctionality. Prog Hum Geogr 29(5):581–600

Schläpfer F (2008) Determinants of voter support for a Five-Year Ban on the cultivation of genetically modified crops in Switzerland. J Agric Econ 59(3):421–435

Schneider MR, Paunescu M (2012) Changing varieties of capitalism and revealed comparative advantages from 1990 to 2005: a test of the Hall and Soskice claims. Socio Econ Rev 10(4):731–754

Sensi A (2016) Agriculture and environment. http://ec.europa.eu/agriculture/envir/report/en/clima_en/report_en.htm (26 Oct 2016)

Soland M, Steimer N, Walter G (2013) Local acceptance of existing biogas plants in Switzerland. Energy Policy 61(4):802–810

Solga H (2014) Education, economic inequality and the promises of the social investment state. Socio Econ Rev 12(2):269–297

Spoerer M (2015) Agricultural protection and support in the European Economic Community, 1962–92: rent-seeking or welfare policy? Eur Rev Econ Hist. https://doi.org/10.1093/ereh/hev001

Steinhaus H (1956) Sur la division des corps matériels en parties. Bull Acad Polon Sci IV(C1.III):801–804

Sulemana IH, James HS Jr (2014) Farmer identity, ethical attitudes and environmental practices. Ecol Econ 58(1):49–61

Tangermann S (2004) Farming support—the truth behind the numbers. OECD Observer 243:38–39

Te Velde H, Aarts N, van Woerkum C (2002) Dealing with ambivalence: farmers' and consumers' perceptions of animal welfare in livestock breeding. J Agric Environ Ethics 15(3):203–219

Tickamyer AR (1983) Rural-urban Influences on legislative power and decision making. Rural Sociol 48(1):133–135

Todaro MP, Stilkind J (1981) City bias and rural neglect: the dilemma of urban development. Population Council, New York

Tyers R, Anderson K (1988) Liberalising OECD agricultural policies in the uruguay round: effects on trade and welfare. J Agric Econ 30(3):197–216

Vesala HT, Vesala KM (2010) Entrepreneurs and producers: identities of Finnish farmers in 2001 and 2006. J Rural Stud 26(1):21–30

Waithaka MM, Thornton PK, Herrero M, Sheperd KD (2006) Bio-economic evaluation of farmers' perceptions of viable farms in Western Kenia. Agric Syst 90(2):243–271

Weible D, Christoph-Schulz I, Salamon P, Zander K (2016) Citizens' perception of modern pig production in Germany: a mixed-method research approach. Br Food J 118(8):2014–2032

Wüstemann H, Mann S, Müller K (2008) Multifunktionalität – von der Wohlfahrtsökonomie zu neuen Ufern. Oekom, München